A-LEVEL

MATHS YEAR 1 AND AS

IN A WEEK

7 DAYS

ROSIE BENTON
SHARON FAULKNER

CONTENTS

DAY 1

Page	Estimated time	Topic	Date	Time taken	Completed
4	40 minutes	Indices and Surds			☐
8	60 minutes	Polynomials			☐
14	60 minutes	Simultaneous Equations and Inequalities			☐
20	60 minutes	Graphs and Transformations			☐

DAY 2

Page	Estimated time	Topic	Date	Time taken	Completed
24	60 minutes	Coordinate Geometry			☐
28	40 minutes	Binomial Expansion			☐
32	60 minutes	Trigonometry 1			☐
36	45 minutes	Trigonometry 2			☐

DAY 3

Page	Estimated time	Topic	Date	Time taken	Completed
40	30 minutes	Exponentials			☐
44	50 minutes	Logarithms			☐
48	45 minutes	Log Rules			☐
52	60 minutes	Using Logarithms			☐

DAY 4

Page	Estimated time	Topic	Date	Time taken	Completed
56	60 minutes	Differentiation 1			☐
60	60 minutes	Differentiation 2			☐
64	40 minutes	Integration 1			☐
68	45 minutes	Integration 2			☐

DAY 5

Page	Estimated time	Topic	Date	Time taken	Completed
74	30 minutes	Modelling, Quantities and Units			☐
78	45 minutes	Kinematics and Graphs			☐
84	50 minutes	Vectors 1			☐
88	50 minutes	Vectors 2			☐

DAY 6

Page	Estimated time	Topic	Date	Time taken	Completed
92	60 minutes	Kinematics 2			☐
96	60 minutes	Forces			☐
102	50 minutes	Proof			☐
106	50 minutes	Measures of Location and Spread			☐

DAY 7

Page	Estimated time	Topic	Date	Time taken	Completed
110	45 minutes	Probability			☐
114	25 minutes	Statistical Sampling			☐
116	50 minutes	Representing Data			☐
120	50 minutes	Statistical Distributions			☐
124	50 minutes	Hypothesis Testing			☐

128 Answers
144 Index

Indices and Surds

Laws of Indices

Indices are expressed as a superscript to a value, in generalised form a^n. If this number, n, is a **positive integer** it represents repeated multiplication.

Example
$4^3 = 4 \times 4 \times 4$

At AS-level recognising, interpreting and manipulating indices is important and it is a basic assumed skill within many other topics. It is also an area where marks are often lost because students either rush it or haven't understood it fully in the first place. Take your time whenever and wherever you meet indices!

It is important to understand the laws, rather than just try to remember them. A short example can be really helpful to confirm the law has been remembered correctly. One is included with each of the index laws below.

$$x^{-a} = \frac{1}{x^a}$$

By considering the patterns of repeat multiplication for positive indices, it is possible to extend this model to include all integer values of n.

Example

$p^{-2} \quad p^{-1} \quad p^0 \quad p^1 \xrightarrow{\times p} p^2 \xrightarrow{\times p} p^3$

$\frac{1}{p^2} \xleftarrow{\div p} \frac{1}{p} \xleftarrow{\div p} \frac{p}{p} = 1 \xleftarrow{\div p} p \xleftarrow{\div p} p \times p \xleftarrow{\div p} p \times p \times p$

$$x^a \times x^b = x^{a+b}$$

The important thing here is that the base numbers (in this case represented by the letter x) must be the same as each other. Sometimes it is possible to change the base to make it the same so that you can simplify using this law. More on that in a moment.

A simple example whilst considering what the indices mean can help with this law.

Example
$2^2 \times 2^3 = 2 \times 2 \times 2 \times 2 \times 2 = 2^5$

$$x^a \div x^b = x^{a-b}$$

Given that $x^a \div x^b = x^a \times \frac{1}{x^b} = x^a \times x^{-b} = x^{a+(-b)}$ it is possible to see where this law of division comes from. Once again it is possible to consider a simple example to double check the law has been remembered and used correctly.

Example

$2^3 \div 2^5 = \frac{\cancel{2} \times \cancel{2} \times \cancel{2}}{\cancel{2} \times \cancel{2} \times \cancel{2} \times 2 \times 2}$
$= \frac{1}{2^2}$
$= 2^{-2}$

Alternatively,
$2^3 \div 2^5 = 2^3 \times \frac{1}{2^5}$
$= 2^3 \times 2^{-5}$
$= 2^{(3+-5)}$
$= 2^{-2}$

$$(x^a)^b = x^{ab}$$

Example

$(2^2)^3 = 2^2 \times 2^2 \times 2^2$
$= 2 \times 2 \times 2 \times 2 \times 2 \times 2$
$= 2^6$
$= 2^{2 \times 3}$

$$x^{\frac{a}{b}} = \sqrt[b]{x^a} = \left(\sqrt[b]{x}\right)^a$$

Example

$x^{\frac{1}{2}} \times x^{\frac{1}{2}} = x^{\frac{1}{2}+\frac{1}{2}} = x \quad \text{so } x^{\frac{1}{2}} = \sqrt{x}$

Extending this further,

$x^{\frac{1}{3}} \times x^{\frac{1}{3}} \times x^{\frac{1}{3}} = x = x^{\frac{1}{3}+\frac{1}{3}+\frac{1}{3}} \quad \text{so } x^{\frac{1}{3}} = \sqrt[3]{x}$

If the numerator has a value other than 1, the root is then raised to the associated integer power.

Example
$25^{\frac{3}{2}} = \left(\sqrt[2]{25}\right)^3 = 5^3 = 125$

It is generally advisable to take the root using the denominator first. The same result will happen either way but it often makes the calculations simpler. Otherwise the intermediate step above would have been $\sqrt{15625}$.

Applying the Laws and Manipulating Indices

The previous rules can be used to interpret and to simplify an expression, whether algebraic or numerical.

Matching Bases

In all the laws, the base of the number has to be consistent in order to be able to apply the law. If the bases don't match it may be possible to manipulate them until they do.

Example

Express $18^2 \times 3^{\frac{2}{3}} \div 4$ in the form 3^p where p is a rational number to be found.

If unsure about how to start, consider using prime factorisation of numbers. This should show where common factors are. In this case, the question suggests the final form should have 3 as the base, so finding a way to write the other bases in terms of 3 could also be your starting point.

$$18^2 \times 3^{\frac{2}{3}} \div 4 = (2 \times 3^2)^2 \times 3^{\frac{2}{3}} \div 2^2$$
$$= 2^2 \times 3^4 \times 3^{\frac{2}{3}} \div 2^2$$
$$= 3^{4+\frac{2}{3}}$$
$$= 3^{\frac{14}{3}}$$

Note: Indices are left as improper fractions rather than converted to mixed numbers.

Evaluating Expressions in the Form $x^{-\frac{b}{a}}$

It is best to think about these types of questions as three distinct steps.

Step 1 Is there a negative? If yes, use the reciprocal of x. (This is $\frac{1}{x}$ but if x is already a fraction it effectively gets flipped upside-down.)

Step 2 Is there a denominator? If yes, consider taking the associated root next.

Step 3 Raise the whole of your base to the power.

All exams now allow calculators. Use these to check your answer but be aware for questions that prompt for clear working or exact answers.

Example

Evaluate $\left(\frac{9}{16}\right)^{-\left(\frac{3}{4}\right)}$, giving your answer in the form $\frac{a}{\sqrt{b}}$.

The negative in the power can be 'used' to flip the fraction upside-down:

$$\left(\frac{9}{16}\right)^{-\left(\frac{3}{4}\right)} = \left(\frac{16}{9}\right)^{\left(\frac{3}{4}\right)}$$

The fourth root can be 'invoked' to find the fourth root of both the numerator and the denominator:

$$= \left(\frac{\sqrt[4]{16}}{\sqrt[4]{9}}\right)^3 = \left(\frac{2}{\sqrt{3}}\right)^3$$

Both numerator and denominator are cubed:

$$= \frac{2^3}{\sqrt{3}^3} = \frac{8}{\sqrt{3^3}} = \frac{8}{\sqrt{27}}$$

Understanding how the rules work and what the different elements in an index mean will help to answer questions that are not simple to plug into a calculator. A calculator will always give the simplest form so exam-style questions will include algebraic terms in the powers, have answers that aren't fully simplified (questions may specify the form of the answer, as above), or demand to see the individual steps taken when answering the question. If a question says 'show that', detailed mathematical reasoning must be shown and the result from a calculator will not gain the marks.

Surds

A surd is a representation of a number using a root sign (most often the square root but it could be any root). Surds are used to accurately represent irrational numbers (i.e. numbers that can't be written accurately as a fraction). If a question asks for the exact answer, it will involve surds or π or e. (There are more irrational numbers but none as significant as these, or needed at AS-level.)

Surds are just numbers and obey all the arithmetic and algebraic rules. They should be written clearly, with any rational multiplier before the surd. This helps to make it clear what is within the root and what isn't.

There are certain mathematical conventions to adhere to when dealing with surds. As ever, the aim is to simplify a number or expression as much as possible. Mathematically the answer to the previous example, $\frac{8}{\sqrt{27}}$, is not yet complete. Type it into a calculator and it might suggest the number is $\frac{8\sqrt{3}}{9}$. There are two key steps in this manipulation.

1. Simplify a Surd

If a number under a square root is a square number, then square root it.

> **Example**
> $\sqrt{9} = 3$

If a number under a square root has a factor that is a square number, it can be simplified.

> **Example**
> $\sqrt{18} = \sqrt{9} \times \sqrt{2} = 3\sqrt{2}$

The same would apply to cube, fourth, etc., roots too.

> **Example**
> $\sqrt[5]{96} = \sqrt[5]{32} \times \sqrt[5]{3} = 2\sqrt[5]{3}$

Familiarity with square and cube numbers is helpful here.

2. Rationalise the Denominator

Mathematically the convention is that, for a final answer, all surds should be represented in the numerator, and none in the denominator. The method used to rectify this problem is known as 'rationalising the denominator'.

Rationalising Simple Denominators (the case where the denominator is a single surd, i.e. $\sqrt{a}$)

To rationalise a surd, multiply it by itself (in the case of square roots), or itself twice in the case of cube roots, etc.

To multiply only the denominator by a value (other than 1) would change the value of the number (or the expression) so must not be done.

In the case where the original number is $\frac{q}{a\sqrt{b}}$, where both a and b are rational, it can be multiplied by $\frac{\sqrt{b}}{\sqrt{b}} = 1$:

$$\frac{q}{a\sqrt{b}} \times \frac{\sqrt{b}}{\sqrt{b}} = \frac{q \times \sqrt{b}}{a\sqrt{b} \times \sqrt{b}} = \frac{q\sqrt{b}}{ab}$$

> **Example**
> $\frac{4\sqrt{3}}{\sqrt{2}} = \frac{4\sqrt{3}}{\sqrt{2}} \times \frac{\sqrt{2}}{\sqrt{2}} = \frac{4\sqrt{3 \times 2}}{2} = \frac{4\sqrt{6}}{2} = 2\sqrt{6}$

Rationalising More Complex Denominators

When the denominator contains the sum or difference of two numbers, at least one of which is a surd, the difference of two squares is used:

$$x^2 - y^2 = (x + y)(x - y)$$

In effect, both squares become rational, and there are no x or y terms.

If the denominator of a fraction is $(a + b)$, where either a, b or both can be irrational, the denominator needs to be multiplied by $(a - b)$, so the numerator must be multiplied by the same.

$$\frac{q}{a\sqrt{b} + c\sqrt{d}} \times \frac{a\sqrt{b} - c\sqrt{d}}{a\sqrt{b} - c\sqrt{d}} = \frac{q(a\sqrt{b} - c\sqrt{d})}{a^2b - c^2d}$$

> **Example**
> Rationalise the denominator $\frac{4}{\sqrt{2} - 3\sqrt{5}}$
>
> $$\frac{4}{\sqrt{2} - 3\sqrt{5}} \times \frac{\sqrt{2} + 3\sqrt{5}}{\sqrt{2} + 3\sqrt{5}} = \frac{4(\sqrt{2} + 3\sqrt{5})}{2 - 9 \times 5}$$
> $$= \frac{4\sqrt{2} + 12\sqrt{5}}{2 - 45}$$
> $$= -\frac{4\sqrt{2} + 12\sqrt{5}}{43}$$

Links to Other Concepts

- Solving equations
- Logs and exponentials
- Integration/differentiation
- Transformation of graphs
- Coordinate geometry
- Pythagoras' Theorem

SUMMARY

Positive integer indices represent repeated multiplications of the base number.

- $x^{-a} = \frac{1}{x^a}$
- $(x^a)^b = x^{ab}$
- $x^a \times x^b = x^{a+b}$
- $x^{\frac{a}{b}} = \sqrt[b]{x^a} = \left(\sqrt[b]{x}\right)^a$
- $x^a \div x^b = x^{a-b}$

Surds represent a set of irrational numbers. They are often needed to write a number 'exactly' (i.e. without rounding).

$\sqrt{a} \times \sqrt{a} = a$, so the square root of a number is the number that when multiplied by itself gives the value of the number under the root sign.

- **Surds can be simplified by extracting square factors from inside a square root (or cube numbers from inside a cube root).**

Conventionally, surds should be represented in the numerator of a fraction. To rationalise the denominator, multiply the denominator and the numerator by the same value.

- **If the denominator is $a + \sqrt{c}$ then the multiplier is $a - \sqrt{c}$.**

Converting from index form can help to understand numbers better. Converting into index form means numbers can be combined and then differentiated or integrated more easily.

QUICK TEST

1. If $\sqrt[3]{q} = q^a$, state the value of a.
2. Simplify $x^3 + x(2x^2 + x) - \frac{x^4 + 3x^3}{x}$
3. Express $\left(\frac{4}{x}\right)^{-\left(\frac{1}{2}\right)}$ without the use of index notation.
4. Rationalise the denominator of $\frac{2}{3\sqrt{6}}$, giving the answer in fully simplified form. Use a calculator to check the answer.
5. Rationalise the denominator of this expression: $\frac{\sqrt{2}}{x - 4\sqrt{2}}$
6. When attempting to integrate $\int \sqrt[3]{x^2} + \frac{2}{x^4}\,dx$, which of the following could be the first step?

   A $\int x^{\left(\frac{3}{2}\right)} + 2x^{-4}dx$

   B $\int x^{\left(\frac{3}{2}\right)} - 2x^{4}dx$

   C $\int x^{\left(\frac{2}{3}\right)} + 2x^{-4}dx$

   D $\int x^{\left(\frac{2}{3}\right)} + 2x^{\left(\frac{1}{4}\right)}dx$
7. Express $\frac{1}{\sqrt{8}}$ in the form a^b, where a is a prime number and b is rational.
8. Write $(x\sqrt{3} + 4)(3x - \sqrt{27})$ in the form $a\sqrt{3} + bx$, where a and b are expressions in terms of x to be found.

PRACTICE QUESTIONS

1. ABC and DEF are similar triangles.

   Express the scale factor from ABC to DEF in the form $\frac{a\sqrt{52}}{b}$ **[4 marks]**

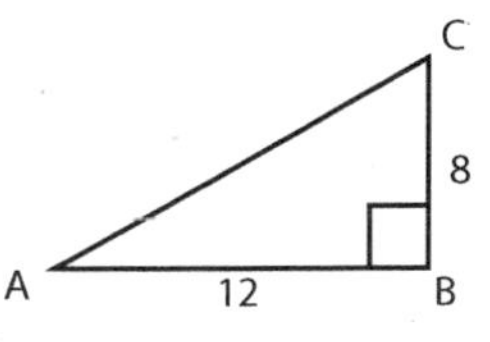

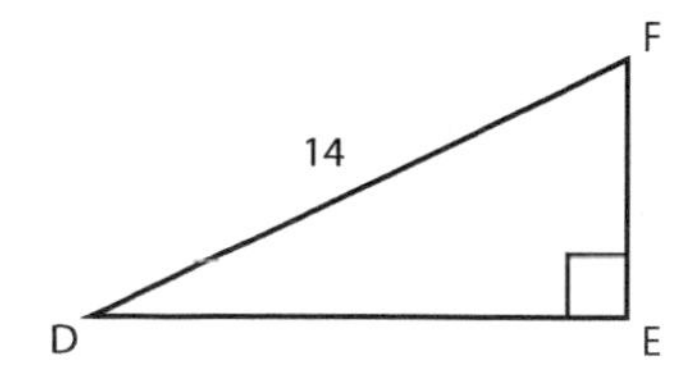

2. Given that $7\sqrt{2}x^2 + a\sqrt{8}x + \frac{16}{\sqrt{b}} = 2^{\frac{1}{2}}(cx^2 + 10x + 2)$, find the values of a, b and c. **[6 marks]**
3. a) Rationalise the denominator in the expression $\frac{4\sqrt{5}}{a+\sqrt{5}}$, where a is a constant. **[2 marks]**

   b) Write the expression $27^{\frac{5}{2}} - 3^{-\frac{3}{2}} + 3^{\frac{1}{2}}$ in the form $(3^a - 3^b + 1)\sqrt{3}$, where a and b are integers to be found. **[6 marks]**

Polynomials

Polynomials

A polynomial is an expression that represents the sum of a set of terms. 'Poly' means multi or many, and 'nomial' means terms. Terms are the individual bits of an algebraic expression separated from each other by + or –.

Simplifying Polynomials

'Like' terms have the same power of x. They can be added or subtracted to simplify the polynomial.

Example

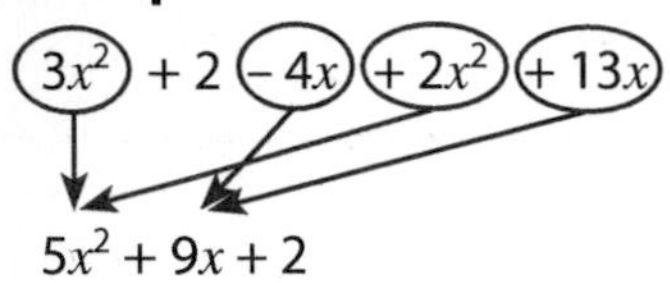

The convention is to write the polynomial in decreasing powers of x.

Example
$3x^3 + x^2 - 4x + 2$
This is a polynomial of order 3, as the highest power of x is 3.

Highest power 2 is also known as a quadratic and highest power 3 a cubic.

Numbers can be expressed as a product of their factors. Polynomials can also be written as a product of factors, and these factors can be polynomial or binomial. A binomial is linear with two terms, e.g. $(2x + 2)$. It is important to be able to manipulate a polynomial to express it both in full form but also as a product of factors.

The full form is useful as it makes it easy to differentiate or integrate. In factorised form it is easy to find roots of the polynomial, and the factors are clear.

Expanding Brackets

The factors expressed within brackets will expand out to show the full polynomial.

Example
Expand $(2x + 3)(x - 1)(3x + 2)$.

Using whichever method you are most comfortable with, start by multiplying out a pair of brackets:

$(2x + 3)(x - 1)(3x + 2)$

Watch out for negatives!

Method 1:

(first) **F:** $2x \times x = 2x^2$

(outside) **O:** $2x \times -1 = -2x$

(inside) **I:** $3 \times x = 3x$

(last) **L:** $3 \times -1 = -3$

$2x^2 + x - 3$

Method 2:

$(2x + 3)(x - 1)$

$2x^2 - 2x + 3x - 3$

$2x^2 + x - 3$

Method 3:

	$+2x$	$+3$
$+x$	$2x^2$	$+3x$
-1	$-2x$	-3

$2x^2 + x - 3$

The next step is to multiply through the quadratic found above with the next bracket. Again, any method is acceptable.

$(2x^2 + x - 3)(3x + 2)$

$6x^3 + 4x^2 + 3x^2 + 2x - 9x - 6$

And simplify:

$6x^3 + 7x^2 - 7x - 6$

Factorising Polynomials

There are a number of approaches that can be used in factorising polynomials. These include: comparing coefficients; polynomial division; by inspection.

The simplest factorisation is when there is a common factor in all terms in the polynomial. In this case, place that value outside the bracket with each term inside the bracket divided by the factor.

Example
Express $6x^4 + 4x^2 - 10x$ in the form $f(x)(ax^3 + bx^2 + cx + d)$.

In this example there is a common factor of $2x$.

$6x^4 + 4x^2 - 10x = 2x(3x^3 + 2x - 5)$

Factor Theorem

The factor theorem is used to identify a factor of a polynomial.

If $f(x) = (x + b)g(x)$, then substituting in $x = -b$ would give a result of $f(-b) = (-b + b)g(x) = 0g(x) = 0$.

The factor theorem states that if $f(a) = 0$ then $(x - a)$ will be a factor of the expression $f(x)$.

Example
Show that $(x - 1)$ is a factor of the expression $3x^3 + 2x - 5$.

$f(1) = 3 \times 1^3 + 2 \times 1 - 5 = 0$

$\therefore (x - 1)$ is a factor.

Simple Algebraic Division

Having identified a factor ($f(x) = (x + b)g(x)$), algebraic division is one method to find the function $g(x)$.

Example

$3x^3 \div x$ → $3x^2$; $3x^2 \div x$ → $3x$; $5x \div x$ → 5

$$
\begin{array}{r|l}
 & 3x^2 + 3x + 5 \\
x - 1 & 3x^3 + 0x^2 + 2x - 5 \\
3x^2 \times (x-1) \rightarrow & -(3x^3 - 3x^2) \\
 & 0 + 3x^2 + 2x \\
3x \times (x-1) \rightarrow & -(3x^2 - 3x) \\
 & 5x - 5 \\
5 \times (x-1) \rightarrow & -(5x - 5) \\
 & 0
\end{array}
$$

Including any terms with coefficient 0 is important.

Take extra care with negatives.

Comparing Coefficients

An alternative method to algebraic division is comparing coefficients.

Example

$$
\begin{aligned}
3x^3 + 2x - 5 &= (x - 1)(ax^2 + bx + c) \\
&= ax^3 + bx^2 - ax^2 + cx - bx - c \\
&= ax^3 + (b - a)x^2 + (c - b)x - c
\end{aligned}
$$

For these two expressions to be equal, the coefficient of each power of x must be equal.

Equating x^3 terms: $a = 3$

Equating x^2 terms: $b - a = 0 \rightarrow b = 3$

Equating x terms: $c - b = 2 \rightarrow c = 5$

Equating x^0 terms: $-c = -5 \rightarrow c = 5$

$f(x) = (x - 1)(3x^2 + 3x + 5)$

At this stage the expression is written as two factors, a linear (or binomial) and a quadratic.

Quadratics

Quadratics are polynomials of order two. They form a family of graphs that have one turning point and are symmetrical about the vertical line through the turning point.

DAY 1

When the coefficient of the x^2 term is positive, the graph forms a ∪ shape (smiley face).

When the coefficient of the x^2 term is negative, the graph forms a ∩ shape (unhappy face).

The constant term (non-x term) gives the y-intercept.

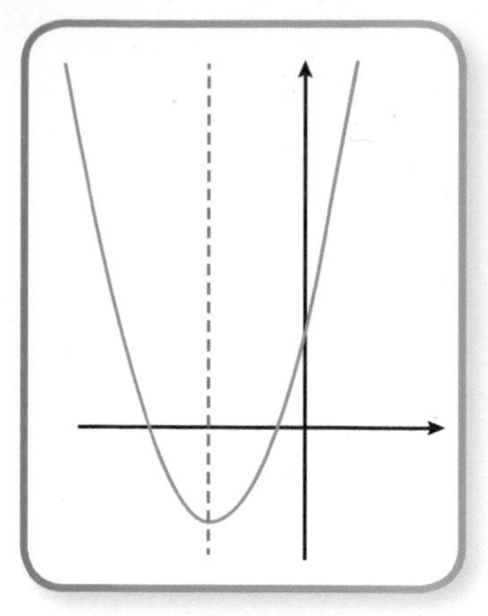

The x-intercepts are known as the roots of the equation, and are what questions are asking for as a solution to an equation.

Solving Quadratics

To solve a quadratic, get all the terms on one side, simplified in an equation equal to zero. Quadratics can have either two distinct real roots, a repeated real root or no real roots. If there are real roots to the equation, it is possible to find them using one of three methods:

1. Factorisation

Further factorisation, again from a multitude of methods, means that a quadratic can be split into two brackets, either of which could be zero to get the desired result.

Example

$2x^2 - 7x - 4 = 0 \rightarrow (2x + 1)(x - 4) = 0$

$2x + 1 = 0 \rightarrow x = -\frac{1}{2}$

$x - 4 = 0 \rightarrow x = 4$

It is always worth looking for factors as it can be the simplest and quickest way to get to the result. If factors aren't obvious then the options are using the quadratic formula or completing the square.

2. Completing the Square

The method of completing the square makes it so all the terms involving an unknown are represented in one place.

Example

$2x^2 - 7x - 4 = 0$

Make it so that the coefficient of the x^2 term is 1, as this is an equation that can be divided by 2 on both sides of the equation. If it was an expression, the factor of 2 that is taken out would be represented outside brackets:

$x^2 - \frac{7}{2}x - 2 = 0$

Place the x in a bracket and add half of the x coefficient (note the negative). This bracket squared expands to produce all the x and x^2 terms, but it has a by-product of $\left(-\frac{7}{4}\right)^2$ so this needs to be subtracted:

$\left(x - \frac{7}{4}\right)^2 - \left(-\frac{7}{4}\right)^2 - 2 = 0$

Rearrange so the constants are together and simplified:

$\left(x - \frac{7}{4}\right)^2 = \frac{81}{16}$

Take the square root of both sides. Both positive and negative roots are required to find all the possible solutions:

$x - \frac{7}{4} = \pm\frac{9}{4}$

$x = \pm\frac{9}{4} + \frac{7}{4} \rightarrow x = 4, x = -\frac{1}{2}$

Completing the square can be used with the generalised quadratic equation ($ax^2 + bx + c$) to find the quadratic formula, which can be applied to any quadratic.

3. The Quadratic Formula

$$x = \frac{-b \pm \sqrt{b^2 - 4ac}}{2a}$$

Example

$2x^2 - 7x - 4 = 0$

First identify terms:

$a = 2, b = -7, c = -4$

Substitute in values:

$x = \frac{7 \pm \sqrt{(-7)^2 - 4 \times 2 \times -4}}{2 \times 2} = \frac{7 \pm \sqrt{81}}{4}$

$\rightarrow x = \frac{7+9}{4} = 4, \quad x = \frac{7-9}{4} = -\frac{1}{2}$

The **discriminant** allows a relatively simple calculation to determine if there are real roots to a quadratic equation.

As mentioned previously, some quadratics have two unique solutions, some have a repeat solution and some have no real solutions. If the quadratic $3x^2 + 3x + 5 = 0$ is used with the quadratic formula, the result is `Math ERROR` or an imaginary number (this depends on the calculator being used – it is an imaginary number if it is written using an i, or possibly a j).

Example

$3x^2 + 3x + 5 = 0 \quad \rightarrow a = 3, b = 3, c = 5$

$x = \frac{-3 \pm \sqrt{3^2 - 4 \times 3 \times 5}}{2 \times 3} = \frac{-3 \pm \sqrt{-51}}{6}$

There are no real roots because the value in the square root is negative. There is no real number that when squared is negative, as a negative multiplied by a negative is positive, and a positive multiplied by a positive is positive.

By separating out the calculation from within the square root of the quadratic equation, it is possible to judge if a number has real roots or not.

- If $b^2 - 4ac > 0$ there are two real distinct roots, represented by the x-intercepts on the graphs.

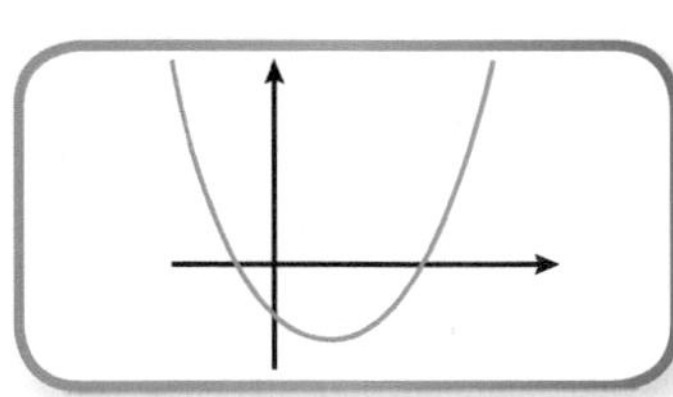

- If $b^2 - 4ac = 0$ there is a repeated real root.

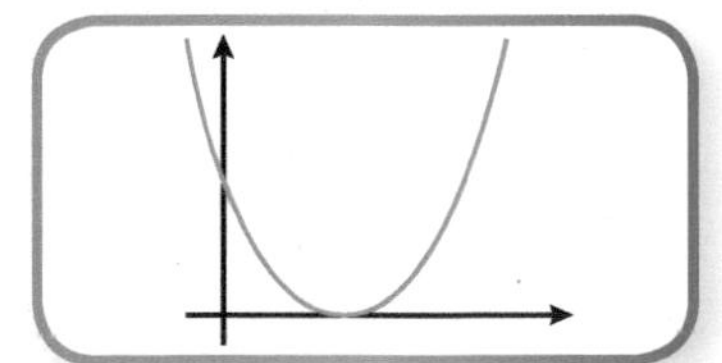

- If $b^2 - 4ac < 0$ there are no real roots to the quadratic equation.

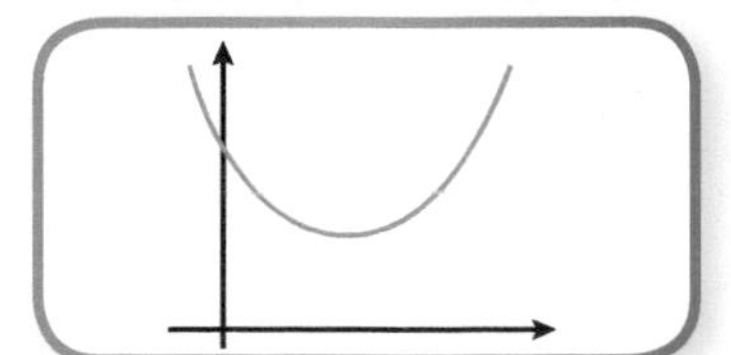

The discriminant can be used to decide if there are roots to a quadratic, or to help find missing terms in a quadratic given the number of real roots.

Example

The equation $px^2 + 3x + 2$ has a repeated root. Find the value of p.

$3^2 - 4 \times 2 \times p = 0$

$8p = 9$

$p = \frac{9}{8}$

Hidden Quadratics

All the quadratics met so far have been shown in terms of x. The unknown value could be represented by any letter or could even be an expression.

Generally look out for one unknown term that is squared. Often people struggle to isolate the unknown term in a quadratic when trying to solve it as a linear equation. Try to arrange it as a quadratic and take it from there. Sometimes it is easier to define an expression with a letter in order to solve the quadratic. This is okay as long as the letter is defined clearly and at the point of getting to an answer it is reinterpreted in terms of the question.

Example

$3\sin^4 x + \sin^2 x = 2\sin^2 x + 1 - 3\sin^4 x$. Find all the possible values of $\sin x$.

At first glance this doesn't necessarily leap out as being a quadratic equation. As there is a $\sin^2 x$ and a $\sin^4 x$, it is possible to make a substitution which will help to reveal the quadratic.

Let $a = \sin^2 x$

$3a^2 + a = 2a + 1 - 3a^2$

For any equation, simplify by collecting like terms.

$6a^2 - a - 1 = 0$

Having revealed the quadratic, solve as such. In this case it does factorise fairly easily.

$(2a - 1)(3a + 1) = 0$

$(2a - 1) = 0 \rightarrow a = \frac{1}{2}$

$(3a + 1) = 0 \rightarrow a = -\frac{1}{3}$

This is solved for a, but since $a = \sin^2 x$, the question is not yet answered. Substitute $\sin^2 x$ back into the results.

$a = -\frac{1}{3} \rightarrow \sin^2 x = -\frac{1}{3}$

There are no valid real results as $\sin^2 x \neq -\frac{1}{3}$

$a = \frac{1}{2} \rightarrow \sin^2 x = \frac{1}{2} \rightarrow \sin x = \frac{\pm\sqrt{2}}{2}$

Any result where there are two variants of the unknown can be treated as a quadratic. For example, $4x^{\frac{4}{3}} + 2x^{\frac{2}{3}} - 7 = 0$ can be treated as a quadratic. A possible substitution is to let $a = x^{\frac{2}{3}}$, which would give a simplified equation of $4a^2 + 2a - 7 = 0$. Substitutions aren't necessary but can be used to make working clearer and easier to follow. Look out for when there are two possible powers and one is double the other.

SUMMARY

- **Polynomials have multiple terms and are written with the highest power of x (or other variable/unknown) first.**
- **To expand a set of brackets, take a pair at a time and make sure each term in the first bracket is multiplied by every term in the second bracket.**
- **To factorise a polynomial, factor theorem can be used to identify a factor:**

  $f(a) = 0$ then $(x - a)$ will be a factor of the expression $f(x)$
- **Having found a factor, algebraic division, comparing coefficients, and inspection are all valid methods of taking the factorising step.**
- **Quadratics are generally factorised by inspection.**
- **Any method of polynomial (including quadratic) factorisation is fine as long as it is clear.**
- **Quadratics can be solved using three key methods:**
  - **factorisation (as above)**
  - **quadratic formula, $x = \frac{-b \pm \sqrt{b^2 - 4ac}}{2a}$**
  - **completing the square.**
- **Using the discriminant,**

  $b^2 - 4ac > 0$, there are two real distinct roots

  $b^2 - 4ac = 0$, there is a repeated real root

  $b^2 - 4ac < 0$, there are no real roots to the quadratic equation.
- **Watch out for quadratics that are hidden in unusual forms.**

Links to Other Concepts

- Solving problems
- Equation of a circle
- Finding roots and solutions
- Inequalities
- Trigonometric equations
- Mechanics and laws of motion
- Statistics
- Simultaneous equations

QUICK TEST

1. Expand and simplify $(3x + 2)(7x - 9)$.
2. Simplify $3b - b^2 + 2b - a^2 + 4b^2$.
3. Expand and simplify $(4x^2 - 2x + 3)(3x - 1)$.
4. Calculate the discriminant for the equation $25x^2 + 30x + 9 = 0$.
5. Match the following graph equations to their graphs.

   A $y = -x^2 + 3x - 4$ **B** $y = 3 + 2x - 2x^2$

   C $y = -x^2 + 2x$ **D** $y = -x^2 + 2x - 1$

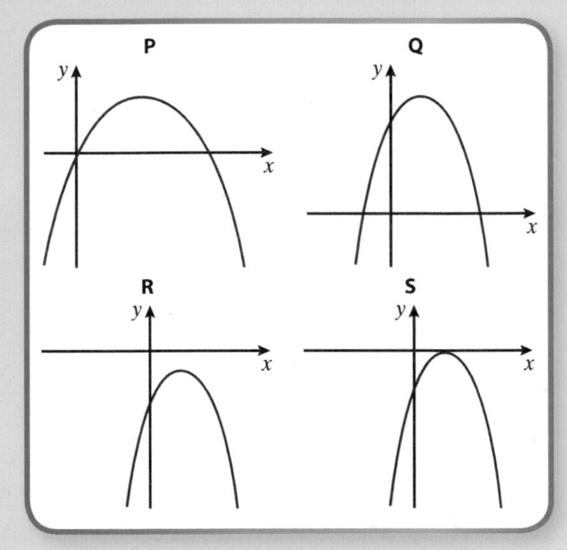

6. State the number of real roots for the following quadratic equations:
   a) $4x^2 + x - 3 = 0$
   b) $8x^2 - 8x + 2 = 0$
   c) $-4x^2 + 8x + 4 = 0$
   d) $4x^2 - x + 3 = 0$
7. The equation $qx^2 + 4.5x + q = 0$ has a repeated root. Find the possible values of q.
8. When completing the square, the equation $x^2 + 4x - 6 = 0$ is written as $(x + p)^2 - q = 0$. Find the values of p and q.

PRACTICE QUESTIONS

1. **a)** Use factor theorem to show that $2x - 1$ is a factor of $12x^3 + 5x - 4$. **[1 mark]**

   b) Divide $12x^3 + 5x - 4$ by $2x - 1$. **[3 marks]**

   c) Show that the resulting quadratic has no real roots. **[2 marks]**

2. **a)** Use the method of completing the square to find the exact solutions to the equation $6x^2 + 8x + 1 = 0$, giving your answer as a single simplified fraction. Show your working clearly. **[5 marks]**

   b) Hence or otherwise find the roots to the equation $10a^{1.2} + 8a^{0.6} - 2 = 4a^{1.2} - 3$, giving answers to 3 significant figures. **[5 marks]**

3. A ball is thrown vertically into the air. On release it is travelling upwards with a speed of $10\,\text{ms}^{-1}$ and is 1.8 m above the ground. Use $10\,\text{ms}^{-2}$ as the downwards acceleration.

   a) Use the equation $s = ut + \frac{1}{2}at^2$, with the values $s = -1.8$, $u = 10$, $a = -10$, $t = ?$, to find the time when the ball hits the ground. **[3 marks]**

   b) Use the equation $s = ut + \frac{1}{2}at^2$ to find the length of time when the ball is 1.8 m above its starting point. **[4 marks]**

Simultaneous Equations and Inequalities

Simple Equations

Equations with a single unknown (only one letter to find the value of) can be solved by collecting like terms together and rearranging to isolate the unknown on one side of the equation. If the unknown is involved in a polynomial, then factor theorem, factorisation and methods for solving quadratics need to be used (factorisation, quadratic formula or completing the square).

Example

Solve the equation $3x - 5 = x + 14$.

$$3x - x = 14 + 5$$
$$2x = 19$$
$$x = 9.5$$

Two Unknowns – Simultaneous Equations

To solve equations with more than one unknown (more than one letter to find the value of), there should be as many independent equations as there are unknowns, i.e. for two unknowns there need to be two equations involving the same variables.

An equation with two simple variables (no index greater than 1) has an infinite number of possible solution pairs that would make the equation true. Graphically this is represented by a straight line.

Example

$3y + 4x = 8$ gives the graph as shown.

Any point on the graph would give a possible pair of values that would present a solution to the equation.

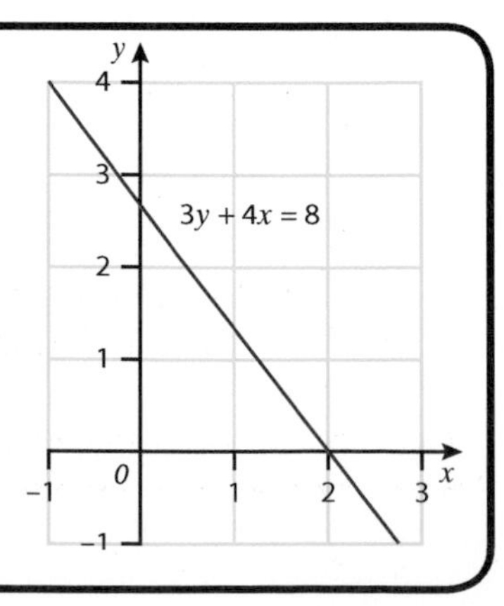

If an independent equation (that is, not just a rearrangement of the same equation) is available, it defines a second relationship between the variables. The intersection of the two graphs gives solutions that fit both equations.

Example

$3y + 4x = 8$ and $6y - 2x + 1 = 0$.

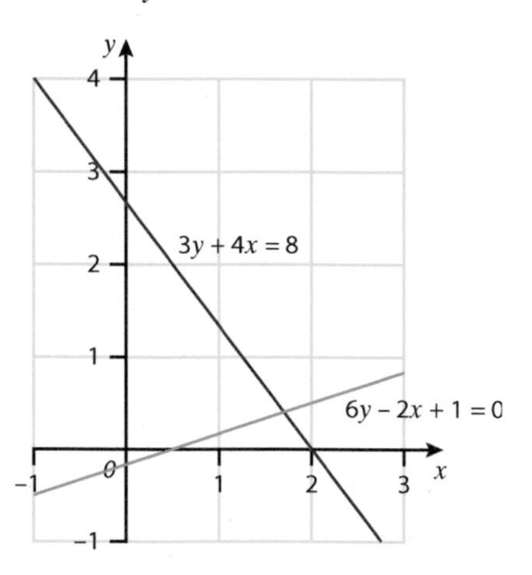

With both these lines represented on the graph, it is possible to see there is a single point that will give a solution to both equations 'simultaneously'.

If both equations are linear, there will be a single point of intersection; just one pair of values that makes both equations true. The only exceptions are if the lines are parallel (no intersections) or identical (infinite solutions).

If one of the equations instead describes a quadratic curve, there may be two distinct results, one repeat result, or possibly even no results. This may be significant when solving problems in coordinate geometry with lines crossing circles, or in mechanics with projectile motion. An example of each of these cases is represented in the form of a quadratic equation and a linear equation in the following graphs.

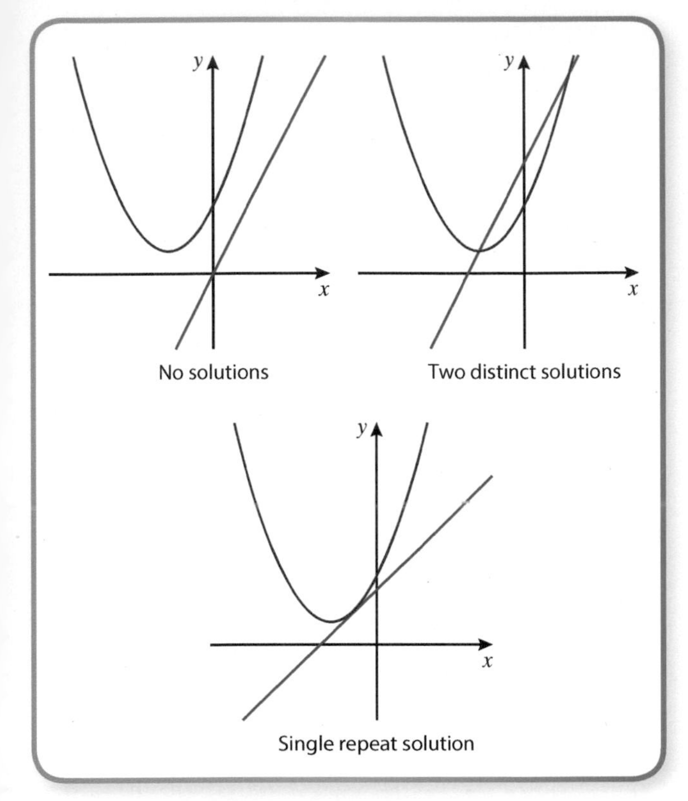

There are two main methods used to solve simultaneous equations algebraically.

Elimination

Elimination works well if there is the same magnitude coefficient for one of the unknowns. In this case the two equations can be added or subtracted in order to 'eliminate' that variable, meaning there is a single equation to work with and only one unknown in it. Elimination can also be used when there is a simple multiplication that can be made to one, or both, equations to create coefficients of equal magnitude.

Magnitude means that the number is the same whether it is positive or negative:

- If both coefficients are the same (both positive or both negative), the two equations should be subtracted (as $ax - ax = 0$ and $-ax - -ax = 0$).
- If the coefficients are negative and positive, the two equations should be added (as $ax + (-ax) = 0$ and $-ax + ax = 0$).

Example

$6x + 2y - 4 = 0$ [1]

$2x - y + 1 = 0$ [2]

In this case, none of the comparable coefficients are of equal magnitude to start with; but if equation [2] was multiplied by 2, the y-coefficients would be of equal magnitude. If equation [2] was multiplied by 3, the x-coefficients would be of equal magnitude. If there are two options, try to select the one that is simpler, but it doesn't matter since the results will be the same either way.

$6x + 2y - 4 = 0$ [1]

$2x - y + 1 = 0$ [2]

[2] $\times 2 \rightarrow 4x - 2y + 2 = 0$ [3]

To eliminate the variables, add equations [1] and [3] (as the signs of the coefficients are different).

$(6+4)x + (2+-2)y - 4 + 2 = 0 \rightarrow 10x - 2 = 0$

Now a variable is eliminated, you can solve the equation by rearranging:

$10x = 2$

$x = 0.2$

Substitute back into either of the equations to solve for y.

Substitute into [2]:

$2 \times 0.2 - y + 1 = 0 \rightarrow 0.4 - y + 1 = 0 \rightarrow y = 1.4$

$x = 0.2$

$y = 1.4$

Substituting back into the other equation helps to check for mistakes.

Check in [1]:

$(6 \times 0.2) + (2 \times 1.4) - 4 = 1.2 + 2.8 - 4 = 0$

Small mistakes often occur when dealing with negative numbers and when doing relatively simple arithmetic, so this is a good way to spot errors and make corrections if needed!

Substitution

Substitution works for all simultaneous equations. Rearrange one equation to find an expression for one variable in terms of the other, then substitute this expression into the other equation. This is the only option when one of the equations is a polynomial or other non-linear graph.

If both equations are linear, it doesn't matter which one is rearranged. Whilst it is possible to rearrange a quadratic to isolate one variable, it is generally easier to rearrange the linear equation and substitute into the quadratic. Again, it doesn't matter which way the question is approached as long as all steps are valid and clearly shown.

Example

$4x-2y-3=0$ and $x+y^2+2y-16=0$. Find the values of x and y, giving each to 3 significant figures.

Rearrange the linear equation to isolate one of the unknown terms:

$2y=4x-3 \rightarrow y=2x-\frac{3}{2}$

Substitute this into the other equation:

$x+\left(2x-\frac{3}{2}\right)^2+2\left(2x-\frac{3}{2}\right)-16=0$

Expand and simplify to form a quadratic and solve:

$x+4x^2-6x+\frac{9}{4}+4x-3-16=0 \rightarrow$

$4x^2-x-16.75=0$

$a=4,\ b=-1,\ c=-16.75$

$x=\frac{1\pm\sqrt{1-4\times4\times(-16.75)}}{8}=\frac{1\pm\sqrt{269}}{8}$

Substitute the fully accurate value (either in surd form or from calculator memory) into either equation (generally the linear will be simpler):

$4\times\frac{1+\sqrt{269}}{8}-2y-3=0 \rightarrow y=\frac{-5+\sqrt{269}}{4}=2.85$ (3 s.f.),

$x=2.18$ (3 s.f.)

$4\times\frac{1-\sqrt{269}}{8}-2y-3=0 \rightarrow y=\frac{-5-\sqrt{269}}{4}=-5.35$ (3 s.f.),

$x=-1.93$ (3 s.f.)

If carrying out a check, remember to use the accurate numbers and not the rounded ones.

Inequalities

Simple Inequalities

An inequality describes a condition where there are a range of possible solutions. They are represented as follows:

- $a<b$ means a is less than b.
- $a>b$ means a is greater than b.
- $a\leqslant b$ means a is less than or equal to b.
- $a\geqslant b$ means a is greater than or equal to b.

Single Variable Inequalities

If there is one variable involved in an inequality, it can be solved like an equation but with some special considerations:

- When rearranging the inequality, consider moving the terms past the fixed inequality symbol.
- The only time the inequality symbol changes (flips from < to >) is when dividing or multiplying by a negative. This is important to bear in mind if the unknown could have a negative value.
- Common sense needs to be applied and considering context may help to make sense of a result.
- Substituting in a value will highlight if there are any mistakes.
- Quadratics and simultaneous equations could be involved in inequalities. Diagrams can help!

Linear Single Unknown Inequality

Example

$4+3x<29-2x$

$4+3x+2x<29$

$5x<29-4$

$x<5$

Alternative method to show the effect of a negative:

$4+3x<29-2x$

$4-29<-2x-3x$

$-25<-5x$

$25>5x$

$5>x$

Using Set Notation to Express Inequalities

Set notation is an alternative way to describe a set of numbers.

- $x \in \{x: x < 5\}$ means x belongs to (is an element of) a set of numbers that are less than 5.
- $x \in \{x: x < 5 \text{ or } x \geqslant 8\}$ means x belongs to a set of numbers that are less than 5, or greater than or equal to 8.
- $x \in \{x: x < 5\} \cup \{x \geqslant 8\}$ means that x is an element of the union of the two sets, so x can be in either.
- $x \in (-\infty, 5) \cup [8, \infty)$ also refers to a union of two sets. The numbers in the brackets represent the limits of the set; a curved bracket is not included in the set but a square bracket is.

Quadratic Inequalities

Quadratic inequalities can be treated in the same way as an equation. Solve them using the methods from pages 10–11 to find the key points, where the inequality changes from being true to false, and then put the inequality signs back in at the end.

Example

$x^2 + x \geqslant 6$

$x^2 + x - 6 \geqslant 0$

$(x+3)(x-2) = 0$

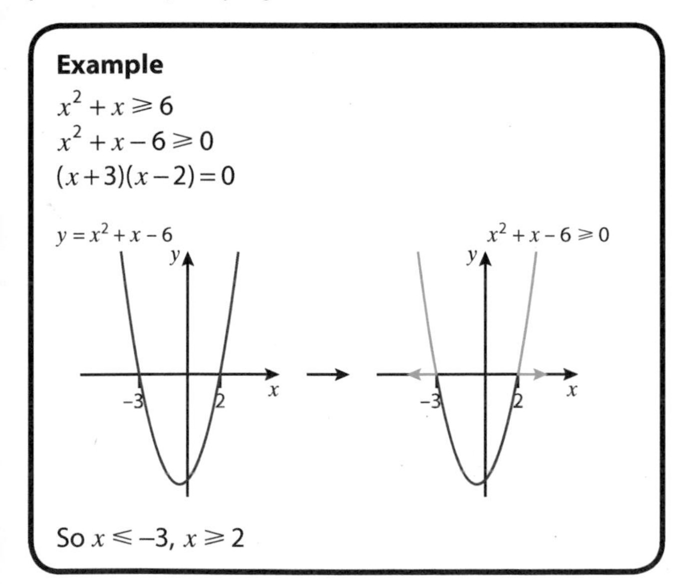

So $x \leqslant -3$, $x \geqslant 2$

If the equation had been $x^2 + x \leqslant 6$, then the final answer would be $-3 \leqslant x \leqslant 2$. This is written together as it is a single section on the graph. The original answer is two separate sections, so is written in two separate inequalities.

If unsure, try putting a value of x into the inequality and see if it is true.

Simultaneous Inequalities and Diagrams

You could be asked for the conditions under which two inequalities are both true at the same time.

Example

$4 + 3x < 29 - 2x$ and $x^2 + x \geqslant 6$

Individually, the solutions to each of these were found earlier. To consider them together, a diagram on a number line can be used (or inspection).

A line, drawn parallel to the number line either above or below, is used to represent the values of x that would make the individual inequalities true. A small empty circle is used on the end to represent $<$ and $>$, but a coloured-in circle is used for $\leqslant$ and $\geqslant$. The two inequalities are represented on this number line:

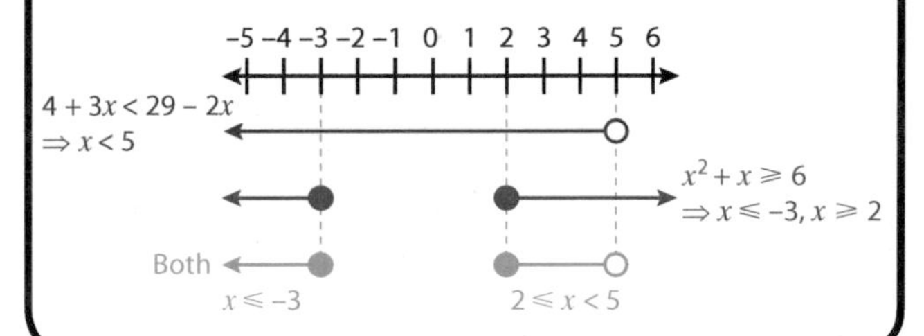

Inequalities with Two Variables

You can represent inequalities involving two unknowns on a graph. Consider the inequality as an equation and use a grid of values or the form $y = mx + c$ to plot the line. The points on each line represent the solutions to the equation. Shading is used to indicate the side of the line which gives solutions to the inequality.

If the line is dotted, it means the points on the line are not included ($<$ or $>$). If the line is solid, it means the points on the line are possible solutions ($\leqslant$ or $\geqslant$).

Example
Shade the region $y+2x-3<0$ on a graph.

Start by considering the line $y+2x-3=0$, which can be rewritten in the form $y=-2x+3$, i.e. a line passing through 3 on the y-axis and with a gradient of –2. A table of results could also be used to draw the line:

x	–1	0	1	2
y	5	3	1	–1

A dotted line is used as the solutions to the equation are not valid solutions to the inequality (<). The area below the graph is shaded.

Quick check – Select any point in the shaded area, e.g. (1, 0), and substitute into the inequality: $0+(2\times 1)-3<0$? Yes, this is true.

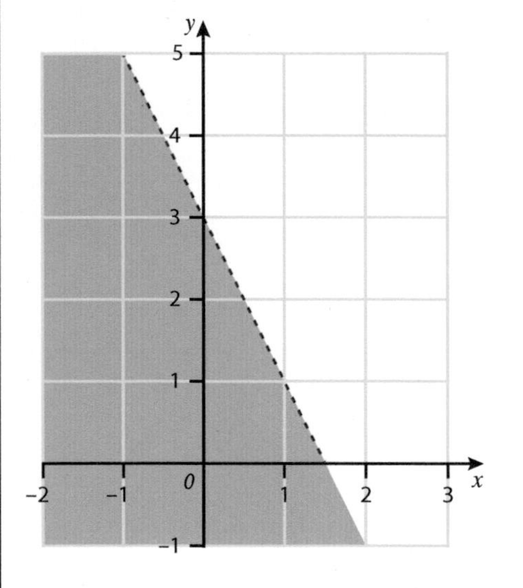

Some questions ask for a region defined by a number of inequalities. Draw each as shown earlier but with only a small amount of shading, or short arrows, to indicate which side of the line is the required region. When all lines are in place, shade the area required.

Example
Shade the region defined by $y\leqslant x$, $x<2$, $y\geqslant\frac{1}{2}$.

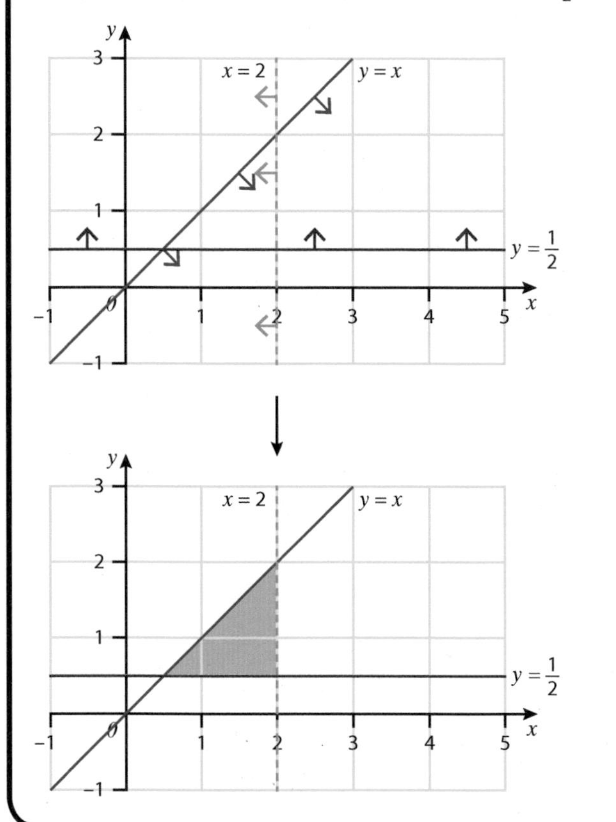

SUMMARY

- **Simultaneous equations can be used to find as many variables as there are independent equations containing them.**
- **The key methods are elimination and substitution.**
- **Graphically, the solutions are represented by the points where two lines cross.**
- **Whilst it is possible to draw the graphs on graphical calculators, it is important to understand what to expect and how to interpret lines.**
- **Inequalities represent a set of numbers.**
- **Set notation, diagrams and inequality symbols can all be used to represent any given inequality.**

Links to Other Concepts

- Comparing coefficients of polynomials
- Ranges and domains of functions – trigonometry
- Problem solving in context (maximisation problems)
- Set notation and statistics
- Quadratics, solving and manipulating – discriminant
- Problems in mechanics and statistics

QUICK TEST

1. a) Find the possible values of t given that $5t + 8 < 2t + 2$.

   b) Write the solution to part **a)** using set notation.

2. Solve the equations $3a + 2b = 2a - b - 3$ and $a - 2b = 10 - a$.

3. Represent the set of numbers such that $x \in (-\infty, 2] \cup [5, 8)$:

   a) on a number line

   b) using the symbols $<$, $>$, $\leqslant$ and/or $\geqslant$.

4. The simultaneous inequality $(ax + 4)(x - a) < 0$ has the solution as represented here:

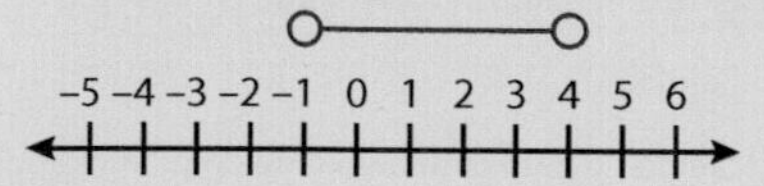

   What is the value of a given that it is a positive integer?

5. $-3 < x \leqslant 4$ and $x \leqslant 1$

   State the possible integer values of x that satisfy both equations.

6. $x^2 - 4x + y^2 + 2y - 20 = 0$ and $x - y - 4 = 0$

   Find the solutions for x and y.

7. Solve the inequality $3q^2 - 15q + 15 \geqslant q^2 - 2q$.

PRACTICE QUESTIONS

1. A straight line has an equation $y = \frac{1}{2}x - 2$ and a curve has the equation $y = x^2 - 2x - 3$.

   a) Use a sketch to show that both intercepts of the two graphs have negative y-coordinates. **[5 marks]**

   b) Show that the x-coordinates satisfy the equation $2x^2 - 5x - 2 = 0$. **[2 marks]**

2. On the grid, shade the region represented by the inequalities $y \geqslant 2x^2 - 11x + 5$ and $y + x < 5$. **[4 marks]**

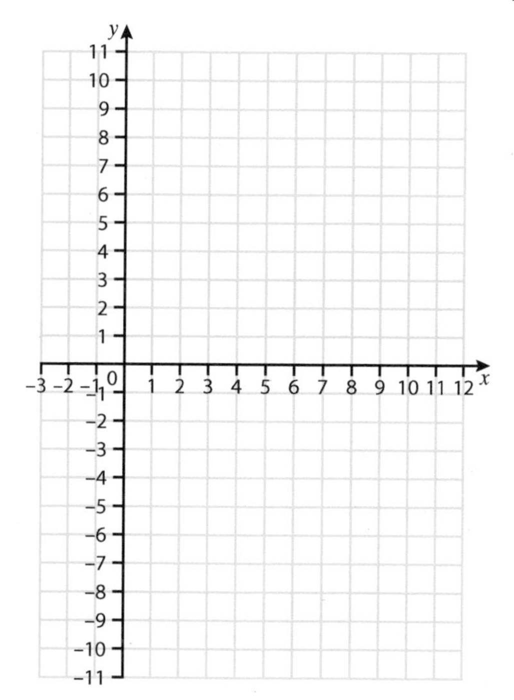

3. Dilys and Tom visit a shop. Tom spends £3.45 and buys two pencils and five pens. Dilys spends £3.30 and buys four pencils and two pens. How much would it cost to buy one pencil and one pen? **[4 marks]**

4. The simultaneous inequalities $ax - b < 0$ and $(x+2)\left(x - \frac{b}{a+5}\right) \geqslant 0$, where a and b are positive integers, are represented on the inequality diagram:

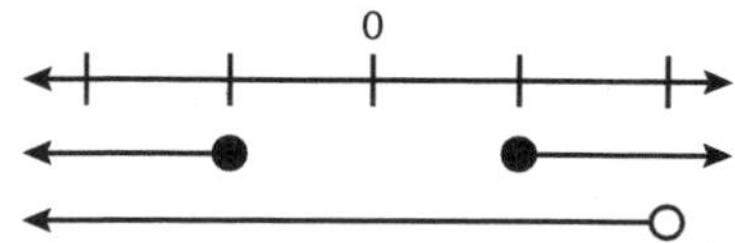

   The solution to the simultaneous inequalities is $x \leqslant -2$ and $2 \leqslant x < 7$. Find the values of a and b. **[7 marks]**

DAY 1 60 Minutes

Graphs and Transformations

Drawing Graphs

Graphs are visual representations of functions. They also show where key elements are, such as **roots**, **asymptotes** and **turning points**. **Sketches** should include key elements, clearly labelled, and the basic shape. They don't need to be on scaled axes. **Drawing** is more accurate and may require the plotting of points. Scaled axes will usually be provided.

Straight-Line Graphs

A **linear function** is represented by a straight line on a graph. The general form is $y = mx + c$. The key features to include on a sketch are:

- the y-intercept, c
- the x-intercept, $-\frac{c}{m}$.

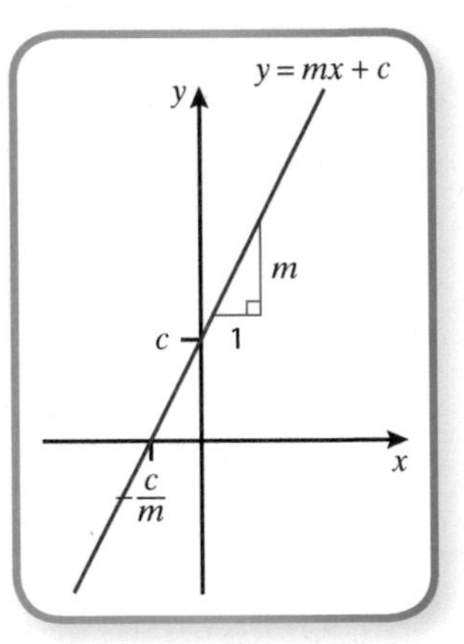

If m is positive, the graph slopes up towards the right /. If m is negative, it slopes down to the right \. The gradient can help to plot the graph; a higher value of m means a steeper line. The gradient doesn't need to be stated on the graph unless asked for.

Quadratics

A quadratic has a general form $y = ax^2 + bx + c$. The key features to include on a sketch are:

- the y-intercept, c
- the x-intercepts, also known as 'roots of the equation', which can be found by setting $y = 0$ and then solving the quadratic
- the maximum or minimum point – the completed square form is an easy way to find this. Also, because the quadratic curve is symmetrical, you can find the midpoint between the x-intercepts, then substitute the value found into the original equation to find the corresponding y-value.

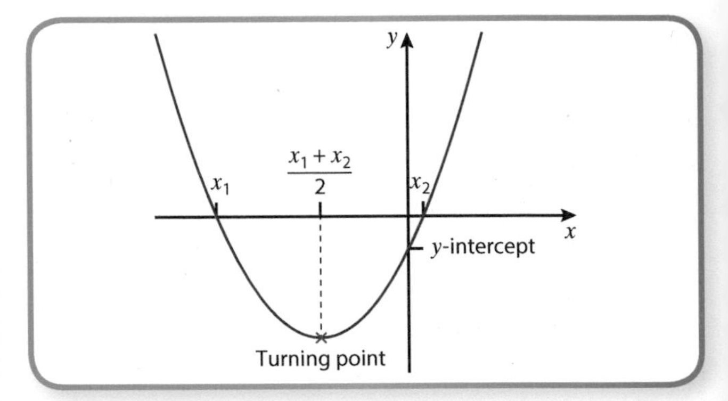

Other Polynomials

Each additional power in a polynomial adds a turning point. The key features to include on a sketch are axis-intercepts and turning points. There are three types of turning point: maxima, minima and points of inflection. In some cases these are in the same place, e.g. $y = x^4$ has all three turning points in the same place. Graphical calculators can help with sketching, but knowing the shapes and possibilities for solutions is useful.

Graph of $y = f(x)$ where:	General shape		Possible solution / roots?		
	a is positive	a is negative			
$y = ax^2 + bx + c$ (quadratic)			No roots	Repeat root	Two distinct roots
$y = ax^3 + bx^2 + cx + d$ (cubic)	Max. Min. Point of inflection		One root	Repeat root Root	Three distinct roots

The **discriminant** can be used with a quadratic to find how many roots there are. The factorised form also gives this information.

Example

$y = x^2 + 4x + 4$ factorises to $y = (x + 2)(x + 2)$.

Both brackets give a result of $x = -2$, so there is a repeat root at -2. It is a positive quadratic with a y-intercept at 4.

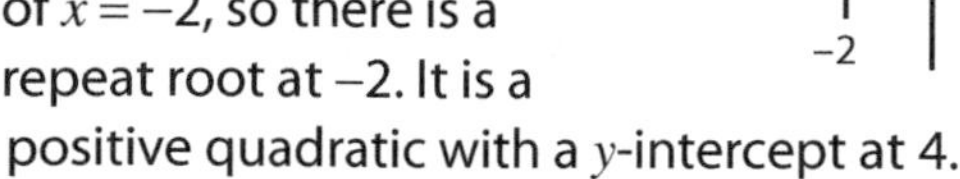

The factorised form will also give this information for other polynomials.

Example

$y = (2x - 3)^2(x + 1)$ would give a curve that is a positive cubic that passes through $x = -1$ then dips down to touch at $x = \frac{3}{2}$ (as this is the 'repeated root').

To find the y-intercept, multiply the constants together: $-3 \times -3 \times 1 = 9$.

If more detail is needed, for example on which side of the y-axis the turning point is, you can use differentiation to locate the turning point precisely. If the exact detail isn't important, then substitute a relatively small value of x into the equation on either side:

$(2 \times 0.1 - 3)^2(0.1 + 1) = 8.62$ (3 s.f.) and
$(2 \times -0.1 - 3)^2(-0.1 + 1) = 9.22$ (3 s.f.)

This shows the graph is decreasing as it passes through the y-axis.

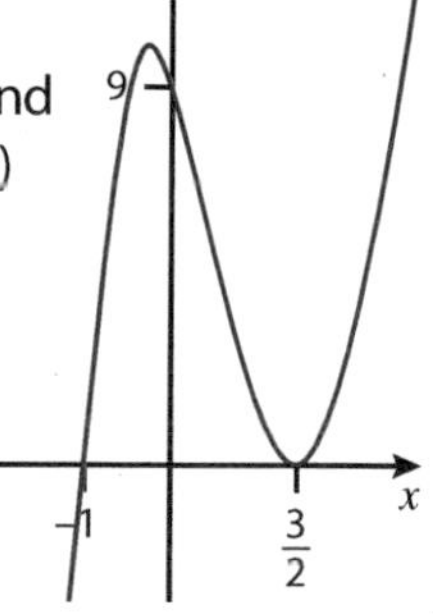

Reciprocal Curves

Reciprocal curves have a function of x in the denominator when expressed as $y = \frac{1}{f(x)}$. Any value divided by 0 is considered to be undefined so **asymptotes** are produced where the x-value makes the denominator equal 0.

Example

$y = \frac{1}{x + 2}$ has a vertical asymptote of $x = -2$.

As the equation can be rearranged to $f(x) = \frac{1}{y}$, the line $y = 0$ also gives an asymptote.

Asymptotes are drawn onto the axes with a dotted line, which represents the value that the curve will approach but never touch.

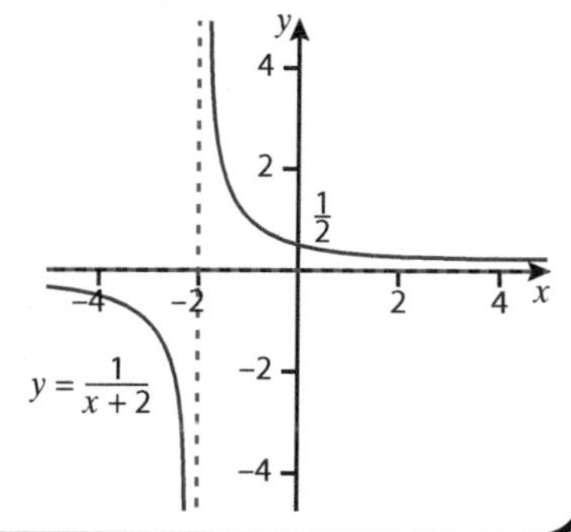

The asymptote means that the graph is discontinuous.

Another key reciprocal graph is $y = \frac{1}{x^2}$. The asymptotes lie on the axes.

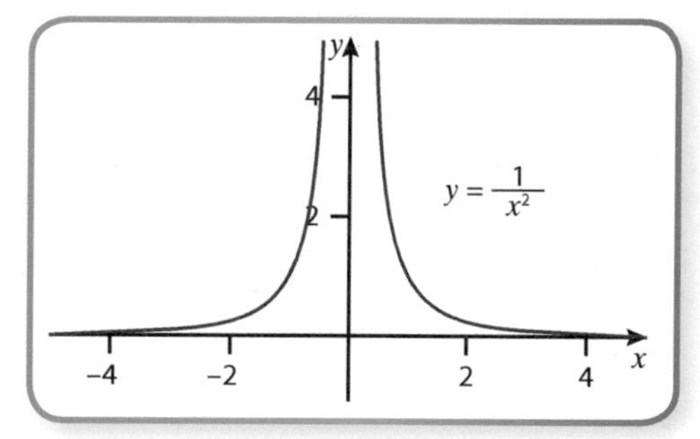

Proportionality

Two values being directly **proportional** will result in a straight-line graph, as $y \propto x$ gives an equation $y = kx$, where k is a constant. If y is proportional to a function of x then the graph will be in the shape represented by $f(x)$, but stretched by a factor of k.

If the variables are inversely proportional, a reciprocal graph is created, as $y \propto \frac{1}{x}$ gives an equation $y = \frac{1}{kx}$ or $y = \frac{k}{x}$.

Exponential Graphs

Graphs of the form $y = a^x$ are exponential.

If $a > 1$, the curve passes through 1 on the y-axis. As x decreases, the curve tends towards the x-axis, which forms an asymptote.

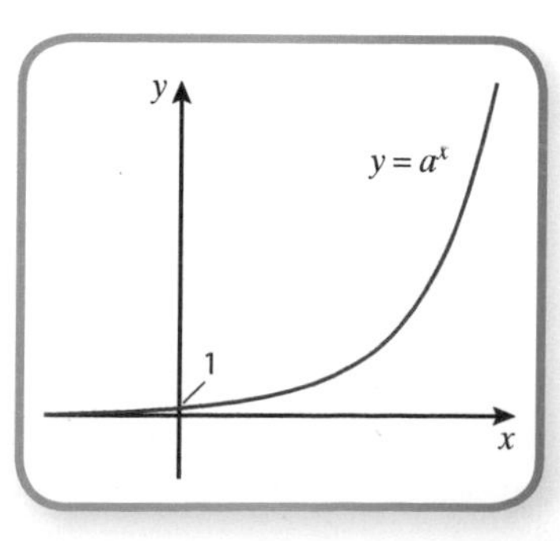

If $0 < a < 1$, the curve is effectively reflected in the y-axis.

DAY 1

Transformations

Using a general function in terms of x, i.e. $y = f(x)$, you can create graphs which have the same traits but have been transformed in some way. The examples here show various transformations of the graph $y = 3x^3 + 9x^2$.

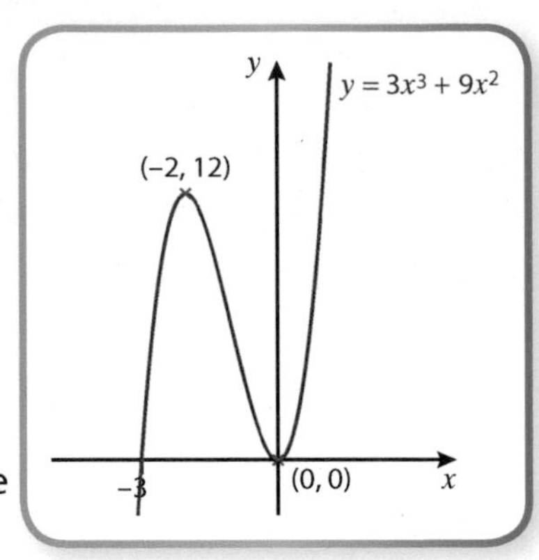

$y = f(x) + a$

The graph is moved up by a, or translated by $\begin{pmatrix} 0 \\ a \end{pmatrix}$. If a is negative, the movement will be downwards.

Example

$y = f(x) + 2$

The graph is moved upwards by 2.

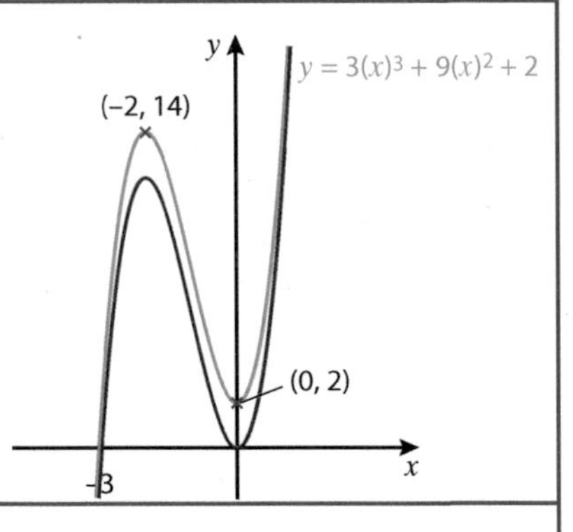

$y = f(x + a)$

The graph is moved left by a, or translated by $\begin{pmatrix} -a \\ 0 \end{pmatrix}$. If a is negative, the movement will be to the right.

Note: This is the same as saying $y = f(x - a)$ is translated by $\begin{pmatrix} a \\ 0 \end{pmatrix}$.

Example

$y = f(x + 1)$

The graph is translated to the left by 1.

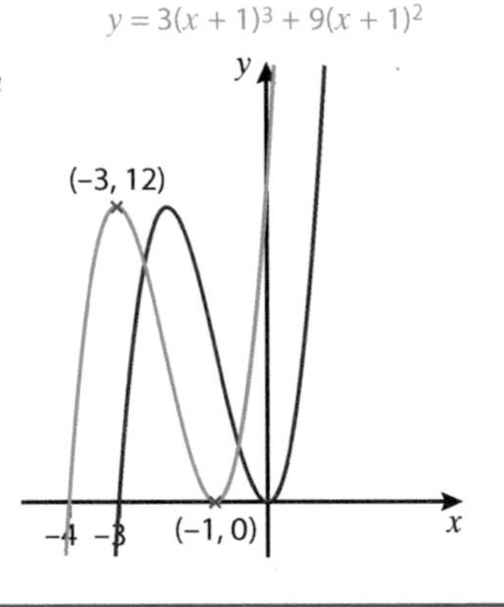

$y = af(x)$, where a is a positive constant

The graph is stretched vertically by a. All x-intercepts stay in the same place. Each y-coordinate is $\times\, a$.

Example

$y = \frac{1}{2}f(x)$

The graph is stretched vertically with a scale factor of $\frac{1}{2}$ (i.e. each coordinate's y-value is halved).

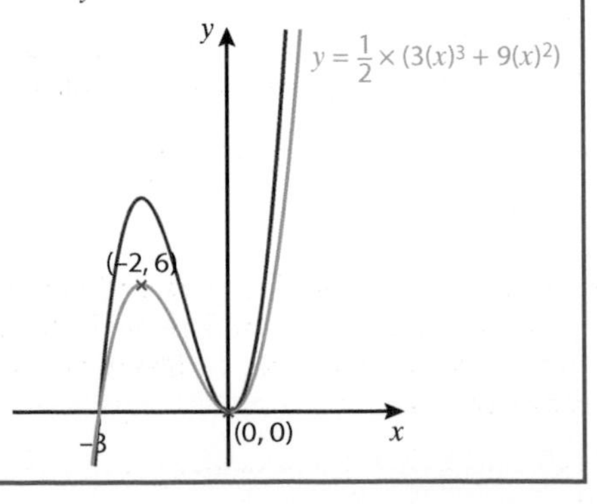

$y = f(ax)$, where a is a positive constant

The graph is stretched horizontally by $\frac{1}{a}$. All y-intercepts stay in the same place. Each x-coordinate is $\times \frac{1}{a}$.

Note: This is the same as saying the translation represented by $y = f\left(\frac{x}{a}\right)$ is a stretch with scale factor a.

Example

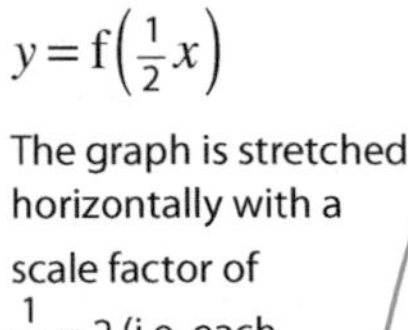

$y = f\left(\frac{1}{2}x\right)$

The graph is stretched horizontally with a scale factor of $\frac{1}{\frac{1}{2}} = 2$ (i.e. each coordinate's x-value is doubled).

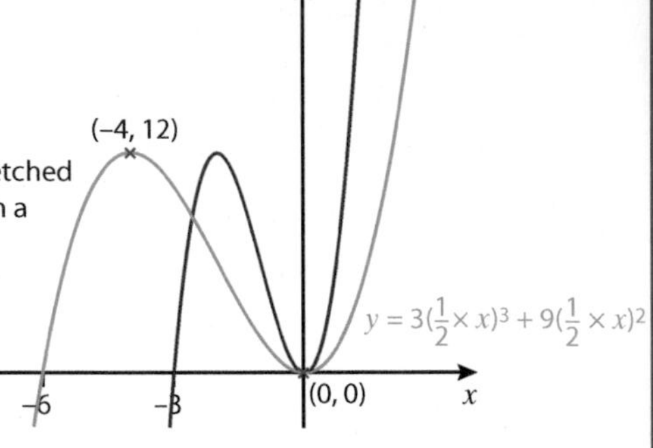

$y = -f(x)$

The graph is reflected in the x-axis.

This transformation is often $y = af(x)$, where a is negative. In that situation it doesn't matter whether it is considered as a stretch then a reflection, or a reflection then a stretch.

Example

$y = -f(x)$

The graph is flipped vertically, i.e. reflected in the x-axis.

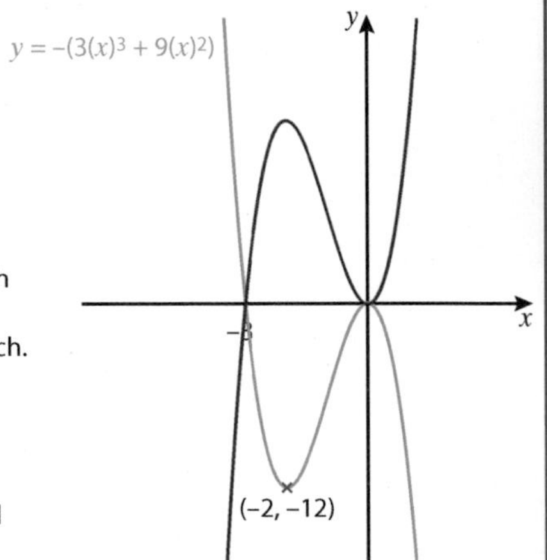

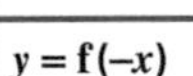

$y = f(-x)$

The graph is reflected in the y-axis.

This transformation is often $y = f(ax)$, where a is negative. In that situation it doesn't matter whether it is considered as a stretch then a reflection, or a reflection then a stretch.

Example

$y = f(-x)$

The graph is flipped horizontally, i.e. reflected in the y-axis.

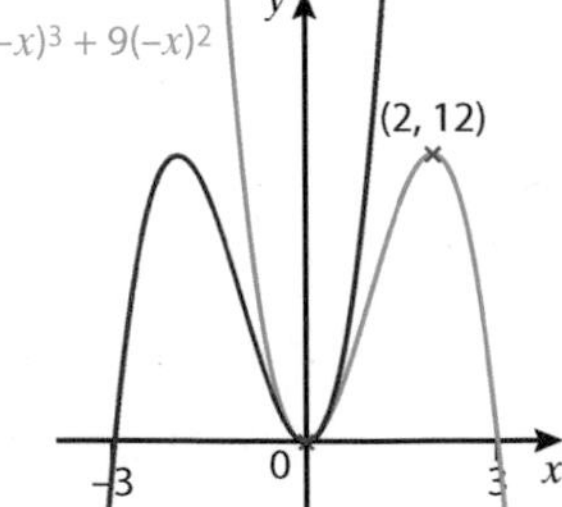

Note that it is possible with certain graphs for two different transformations to give the same result:

- If $f(x) = a^x$ then $f(x + b) = a^{x+b} = a^b \times a^x$. This transformation could be described as either a translation or a stretch.

- If $f(x) = mx$ then $f(x+1) = mx + m = f(x) + m$. This could be considered as either a translation of $\begin{pmatrix} -1 \\ 0 \end{pmatrix}$ or of $\begin{pmatrix} 0 \\ m \end{pmatrix}$.

SUMMARY

- **Key features to include on a graph are intercepts, asymptotes and turning points.**
- **Polynomials form continuous curves. The roots of the equation (x-intercepts) can be found by factorisation or other numerical methods. Factor theorem can help to identify a linear factor for a cubic graph.**
- **If there is a repeat factor, this represents a repeat root on the graph, i.e. the x-axis forms a tangent at the maximum/minimum.**

QUICK TEST

1. Sketch the graph of $y = \frac{1}{x}$.
2. State the equation of the vertical asymptote for the graph $y = \frac{1}{(x+1)^2}$.
3. Sketch the polynomial $y = (x+1)(4-x)(2x-3)$.
4. a) Describe a single transformation that maps $y = 2x + 3$ onto $y = 2x + 5$.

   b) Describe a second transformation that maps $y = 2x + 3$ onto $y = 2x + 5$.
5. A quadratic graph $f(x)$ has its turning point at (3, 4) and the y-intercept at 2. Sketch the graph of $y = 2f(x)$.

PRACTICE QUESTIONS

1. a) Sketch the graph $y = -2x^2 + x$. **[2 marks]**

b) The graph is translated such that it has a single repeated root. Describe the transformation applied to the graph. **[2 marks]**

2. a) Show that the equation $y = x^3 - 3x^2 + 4$ has a factor of $(x+1)$. **[2 marks]**

b) Hence factorise the equation completely. **[4 marks]**

c) Sketch the graph of $y = x^3 - 3x^2 + 4$. **[3 marks]**

d) A translation of $\begin{pmatrix} 2 \\ 0 \end{pmatrix}$ is applied to the graph.

What is the new equation for the graph? Give your answer in the form $y = ax^3 + bx^2 + cx + d$. **[3 marks]**

e) Sketch the new graph, including turning points and axis-intercepts. **[2 marks]**

3. The graph shows the curve $y = f(x)$.

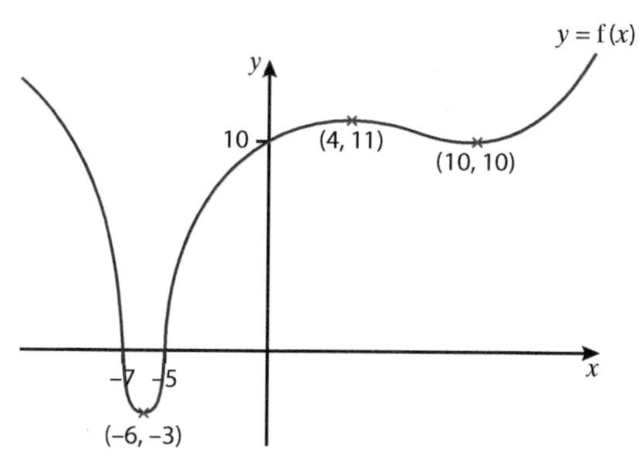

a) What is the y-intercept of the curve $y = f(x) + 2$? **[1 mark]**

b) i) Sketch the graph and state the value of a for which the transformation $\begin{pmatrix} 0 \\ a \end{pmatrix}$ would make the graph $y = f(x)$ pass through the origin. **[3 marks]**

ii) What two other single transformations would make this graph pass through the origin? **[2 marks]**

Links to Other Concepts

- Comparing coefficients of polynomials
- Trigonometric functions
- Differentiation and integration
- Ranges and domains of functions – trigonometry
- Quadratics
- Problems in mechanics and statistics
- Exponentials and logs
- Numerical methods
- Vector notation
- Manipulation of indices
- Algebraic manipulation

Coordinate Geometry

At AS-level, coordinate geometry is in two dimensions (2D) and described using Cartesian **coordinates**. This is a system where points are described, relative to an **origin**, using a grid system or two variables (in 2D) that are set at right angles to each other. A given pair of coordinates gives one point on the plane.

Example
(3, –0.5) means the x-variable equals 3 and the y-variable equals –0.5.

An **equation** gives a relationship between the two variables and describes a set of points. At AS-level these points will form a straight line or a curve.

Straight Lines

A straight-line equation will have two variables (generally x and y). If an equation has x and y variables with index of 1, it is a linear relationship and is represented on a set of axes by a straight line. The equation of a straight line can be written in different forms, the three most common being:

1. $y = mx + c$

This form allows the gradient (m) and the y-intercept (c) to be found without any further calculations.

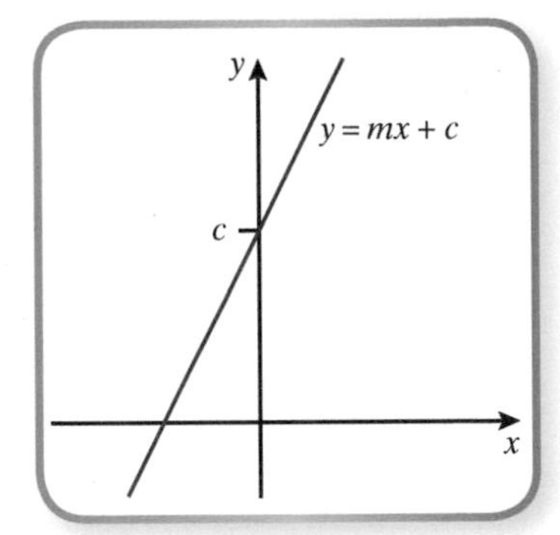

Example
The equation $y = -3x + 0.2$ has a gradient of –3 and a y-intercept of 0.2.

2. $ax + by + c = 0$

This is the form that means that a, b and c can all be integer values. It is often specified in an exam question as the required form. Calculations are required to find the gradient and the y-intercept:

- To find the gradient, use $\frac{-a}{b}$.
- To find the y-intercept, use $\frac{-c}{b}$.

3. $y = a, x = b$

These are simplifications of the form $y = mx + c$, but where the coefficient of one of the variables is equal to 0.

Example
$y = -2$ means that the y-coordinate is always –2, whatever the x-coordinate. It gives a straight line parallel to the x-axis and crossing through –2 on the y-axis. If considered as part of $y = mx + c$, it would be $y = 0x - 2$, so the gradient is 0 (horizontal) and the y-intercept is –2.

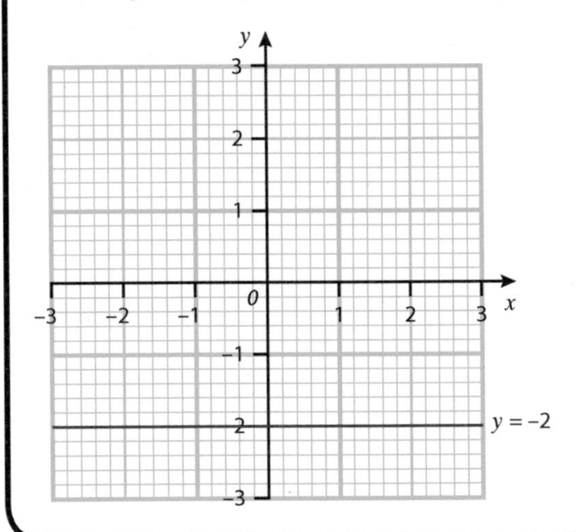

A common mistake is to see the y in the equation ($y = -2$) and expect it to be vertical, i.e. parallel to the y-axis.

Gradients

The gradient of a line is its steepness. It means how many spaces it goes up for each space it goes across. A gradient of $\frac{-1}{2}$ means it goes down half a space for every whole space you move to the right.
It is the change in y divided by the change in x.

A pair of **parallel** lines have the same gradient.

A pair of **perpendicular** lines (i.e. at right angles to each other) have gradients that are **negative reciprocals**. That means the product of their gradients is –1. If line AB is perpendicular to one with a gradient of a, the gradient of AB will be $\frac{-1}{a}$.

Example

Line AB has the equation $2y - x + 6 = 0$ and line CD has the equation $y = -2x + 7$. Are the lines AB and CD parallel, perpendicular or neither? Justify your answer.

Line CD has a gradient of -2. This is easy to read as the equation is already in the form $y = mx + c$.

Line AB can be rearranged to give:

$2y = x - 6$

$y = \frac{1}{2}x - 3 \rightarrow$ the gradient of AB is $\frac{1}{2}$

$\frac{1}{2} \times (-2) = -1$ $\therefore$ the lines are perpendicular as their gradients are negative reciprocals.

Finding the Equation of a Straight Line

$\boldsymbol{y_2 - y_1 = m(x_2 - x_1)}$ can be used to find the equation of a straight line if you are given two points, (x_1, y_1) and (x_2, y_2), that the line passes through or if you are given the gradient (or a way in which to find the gradient) and a point the line passes through. This leads towards an equation in the form $ay + bx + c = 0$.

To find the gradient between two points, use $m = \frac{y_2 - y_1}{x_2 - x_1}$. It doesn't matter which point is (x_1, y_1) and which is (x_2, y_2) as long as they are used within their pairs consistently.

Example

A line goes through two points, P$(-2, 7)$ and Q$(4, 3)$. Find the equation of the line in the form $ax + by + c = 0$, where a, b and c are integer values to be found.

Find the gradient: $m = \frac{(y_2 - y_1)}{(x_2 - x_1)} = \frac{7 - 3}{-2 - 4}$

$= \frac{4}{-6} = \frac{-2}{3}$

To find the equation, substitute the gradient, a known point and (x, y), a general point on the line, into $y_2 - y_1 = m(x_2 - x_1)$. It is generally simplest to substitute the known point as (x_1, y_1).

$y - 7 = \frac{-2}{3}(x - -2)$

This question specifies the form of the equation. Even if it hadn't, this is not yet an acceptable answer as simplification is needed.

$3y - 21 = -2x - 4$

$2x + 3y - 17 = 0$

$a = 2$, $b = 3$ and $c = -17$

Length and Midpoint

Pythagoras' Theorem can be used to find the **length** of the line segment, as the two variables (x and y) are at right angles.

Example

Find the length of the line segment between the points C$(-2, 5)$ and D$(3, -1.5)$ correct to 2 decimal places.

A sketch can help to show what to do:

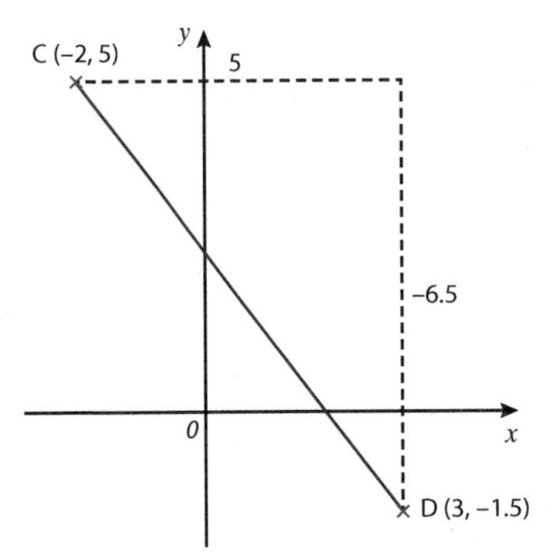

Pythagoras' Theorem states $a^2 + b^2 = c^2$. This question is asking for the length of the hypotenuse so you can use $c = \sqrt{a^2 + b^2}$.

Length $\sqrt{5^2 + (-6.5)^2} = \sqrt{25 + \frac{169}{4}}$

$= \sqrt{\frac{269}{4}} = \frac{\sqrt{269}}{2}$

This is the exact answer. Convert into a decimal to obtain the degree of accuracy asked for.

Length = 8.20 (2 d.p.)

The **midpoint** of a line, where the two end coordinates are known, can be found by adding the x-coordinates and dividing by 2 to find the x-coordinate of the midpoint, and adding the y-coordinates and dividing by 2 to get the y-coordinate. That is, the midpoint of the coordinates (x_1, y_1) and (x_2, y_2) is $\left(\frac{x_1+x_2}{2}, \frac{y_1+y_2}{2}\right)$.

Example

Find the midpoint of the line CD, where C is (–2, 5) and D is (3, –15).

x-coordinate of midpoint $= \frac{-2+3}{2} = \frac{1}{2}$

y-coordinate of midpoint $= \frac{5+-15}{2} = -5$

The midpoint is $\left(\frac{1}{2}, -5\right)$.

Circles

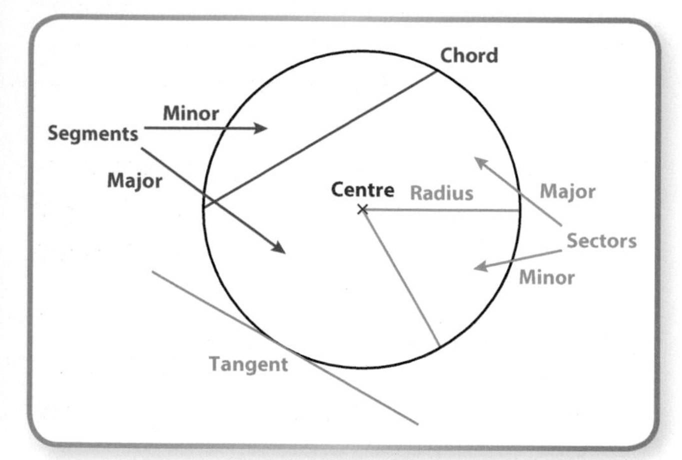

- A tangent to a circle meets the radius at right angles. This means the equations of the two lines are perpendicular so the gradients will be negative reciprocals of each other.
- Lines from either end of a diameter line will meet on the circumference of the circle at right angles.
- Radii are all equal length.
- A radius that meets a chord at a right angle will bisect the chord.

Equations of circles can be written in the form $(x-a)^2+(y-b)^2=r^2$. This form can be used to find the centre of the circle, (a, b), and the radius, r. If an equation is given for a circle but it is not in this form, using the method of **completing the square** will help to do so.

Example

Find the centre and the radius of $y^2+x^2+4y-8x+3=0$.

By grouping the y-terms close together, it is easier to complete the square. Ensure you show your method in clear, short steps.

$y^2+4y+x^2-8x+3=0$

$(y+2)^2-4+(x-4)^2-16+3=0$

For more on how to complete the square, see page 10. To complete the square, both the x^2 and y^2 terms must have a coefficient of 1.

Collect the numbers together and move to the other side of the equation.

$(y+2)^2+(x-4)^2=17$

Centre of the circle is (4, –2); radius is $\sqrt{17}$.

Note: You take the positive square root, as you are talking about a distance.

Links to Other Concepts

- Straight lines describe all directly proportional relationships between simple variables
- Solving simultaneous equations to find the intersection points of two lines
- Transformations and functions
- Straight-line motion in mechanics

SUMMARY

- **A coordinate system describes all the possible points on a plane using two variables, set at right angles to each other. An equation in terms of these variables produces a set of points.**
- **If both variables (x and y) are to the power 1 in an equation, the graph is a straight line.**
- **The gradient of a graph, $m = \frac{y_2 - y_1}{x_2 - x_1}$.**
- **Parallel lines have equal gradients.**
- **If two lines are perpendicular, their gradients are: $m_2 = -\frac{1}{m_1}$ or $m_1 \times m_2 = -1$.**
- **The equation of a straight line is $y = mx + c$ or $y - y_1 = m(x - x_1)$.**
- **To find the midpoint of a line segment, add the end points and divide by 2.**
- **To find the length of a line, use Pythagoras.**
- **The general equation of a circle is $(x-a)^2 + (y-b)^2 = r^2$, where it has centre (a, b) and radius r. To get an equation into this form, use completing the square.**
- **To find the intersection points of two graphs, treat them as simultaneous equations.**

QUICK TEST

1. What is the gradient of the line $y = 5x - 19$?
2. What is the gradient and the length of the line segment between points P(4, –12) and Q(–11, –4)?
3. A circle has centre (–3, –1) and radius 7. Write down the equation of the circle.
4. A tangent meets a circle at point P and has equation $4x - 2y = 3$. What is the gradient of the line joining the centre of the circle and point P?
5. Write an equation for a line that is parallel to $5x + 10y - 4 = 0$ and passes through the origin.
6. Sketch the graph of the equation $(x + 4)^2 + (y - 1)^2 = 25$, showing the centre and where it meets the y-axis.
7. Find the points where the line $y = 2x - 1$ meets the circle $(x + 4)^2 + (y - 1)^2 = 25$.

PRACTICE QUESTIONS

1. A circle passes through the origin, as shown on the graph. A line passes through the centre of the circle, intercepting the circle at points P and Q.

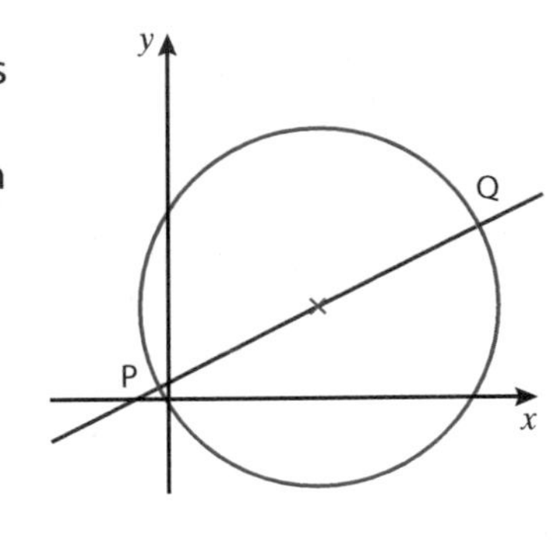

The equation of the line is $y = \frac{1}{2}x + 1$.

   a) Show that the centre of the circle has coordinates $(2a, a + 1)$, where a is a constant. **[2 marks]**

   b) Given that the circle intercepts the y-axis at the origin and again at $y = 10$, find the coordinates of the centre of the circle. **[2 marks]**

   c) Hence find the equation of the circle. **[3 marks]**

   d) What is the exact length of PQ? **[1 mark]**

2. The line that joins points A($2k$, 3) and B(–2, k – 3) has a midpoint (m, m) where m and k are constants.

   a) Find the value of k. **[3 marks]**

   b) Find the exact length of AB. **[1 mark]**

   Line BC is perpendicular to AB.

   c) Find an equation for BC in the form $ax + by + c = 0$, where a, b and c are constant integers. **[4 marks]**

3. Nia has designed a yin yang using semicircles, as shown. Points A, B, C and D all lie on a straight line. Point A is (0, 5) and point B is (12, 0). C is the midpoint of AB and D is the midpoint of BC.

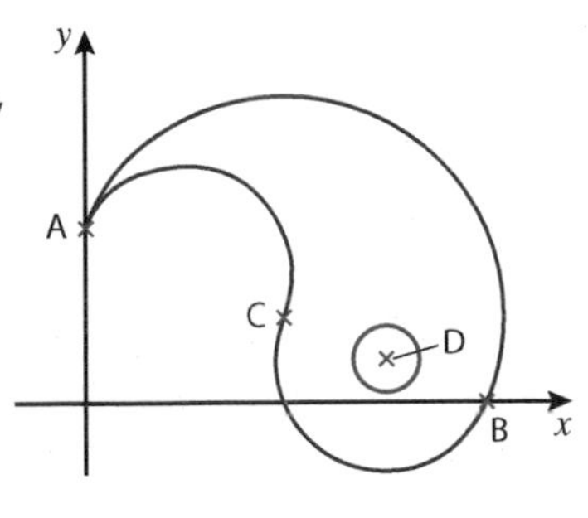

   a) Find the coordinates of point D. **[1 mark]**

   The line $y = kx$, where k is a constant, passes through C and meets a tangent to the semicircle AB at point P.

   b) i) Find k. **[1 mark]**

   ii) Find the coordinates of point P. **[2 marks]**

   iii) Find the equation of the tangent in the form $ax + by + c = 0$, where a, b and c are constant integers. **[3 marks]**

Binomial Expansion

A binomial is an algebraic expression which is the sum, or difference, of two terms. Binomial expansion deals with the expansion of a bracket containing a binomial when it is raised to a power: $(ax+b)^n$

Multiplying Out Brackets

Understanding of the process of binomial expansion comes from the basic concept of how to expand a bracket.

Example

$$(ax+b)^2 = (ax+b) \times (ax+b)$$
$$= (ax)^2 + 2abx + b^2$$

Once the power increases much beyond squaring, it starts to get quite complicated.

Example

$$(ax+b)^3 = (ax+b)(ax+b)(ax+b)$$
$$= (ax+b)\left((ax)^2 + axb + bax + b^2\right)$$
$$= (ax+b)\left((ax)^2 + 2abx + b^2\right)$$
$$= (ax)^3 + 2a^2bx^2 + ab^2x + a^2bx^2 + 2ab^2x + b^3$$
$$= (ax)^3 + 3(ax)^2b + 3(ax)b^2 + b^3$$

By using the patterns that emerge in the coefficients, you can get to this result with simpler working, especially if you only need to find coefficients of particular terms.

Simple Binomial Expansions

For the simple binomial expansions, $(1+x)^n$, where n is a positive integer, the coefficients match the numbers found in Pascal's triangle.

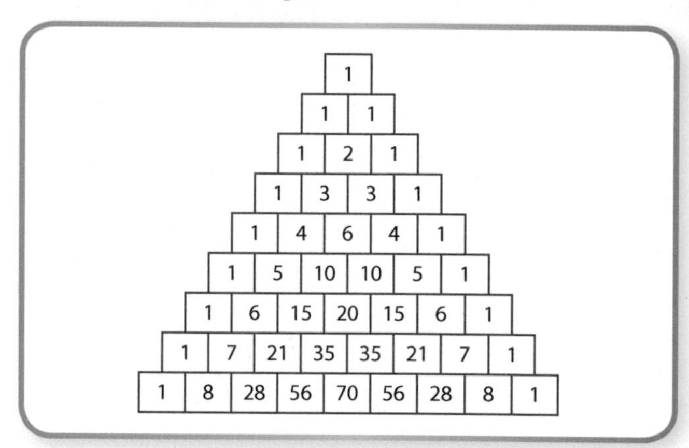

Within Pascal's triangle, each number is calculated by adding the two numbers above it. This can be a quick way to generate the coefficients for relatively low values of n. To find the coefficients for an expansion in the form $(ax+b)^n$, use the numbers in the row that starts 1, n, … .

Example

The coefficients to use for $(ax+b)^4$ are 1, 4, 6, 4, 1.

An alternative way is to use **combinations** to generate each coefficient associated with a term. This can be very efficient if you are only interested in particular terms, or where n is a relatively large number.

Example

Consider $(1+x)^4$.

The expansion will be in the form:

$$p(1)^4(x)^0 + q(1)^3(x)^1 + r(1)^2(x^2) + s(1)^1(x)^3 + t(1)^0(x)^4$$

The values of p, q, r, s and t come from the number of ways in which that term can be generated, i.e.

$(1+x)^4 = (1+x)(1+x)(1+x)(1+x)$

There is one way to get the x^0 term from this expansion.

$(1+x)(1+x)(1+x)(1+x)$

There are four ways to generate the x^1 term in this expansion.

To calculate the number of combinations that produce each term, x^a, we use the theory of combinations. The number of ways each term can be generated is $\frac{n!}{r!(n-r)!}$, which can also be written as $(n, r) = {}_nC_r = {}^nC_r = \binom{n}{r}$, where n is the power and r is the index of the x term being generated.

Example

What is the coefficient of the x^6 term in the expansion of $(1+x)^8$?

$n = 8$ and $r = 6$

The coefficient is $\frac{8!}{6!(8-6)!} = \frac{8 \times 7 \times 6 \times 5 \times 4 \times 3 \times 2 \times 1}{6 \times 5 \times 4 \times 3 \times 2 \times 1 \times (2 \times 1)}$

$= \frac{8 \times 7}{(2 \times 1)} = 28$

This is the seventh value in the $n = 8$ row of Pascal's triangle.

This can be calculated using the factorial button on a calculator or by using the 'combination key' labelled using one of the representations above, generally ${}_nC_r$ or ${}^nC_r$.

Complex Binomial Expansions

For more complex binomial expansions, $(a+bx)^n$, where n is a positive integer and a and b are constants, the coefficients are the product of ${}^nC_r$ (the number of combinations creating the term) and the corresponding powers of a and b:

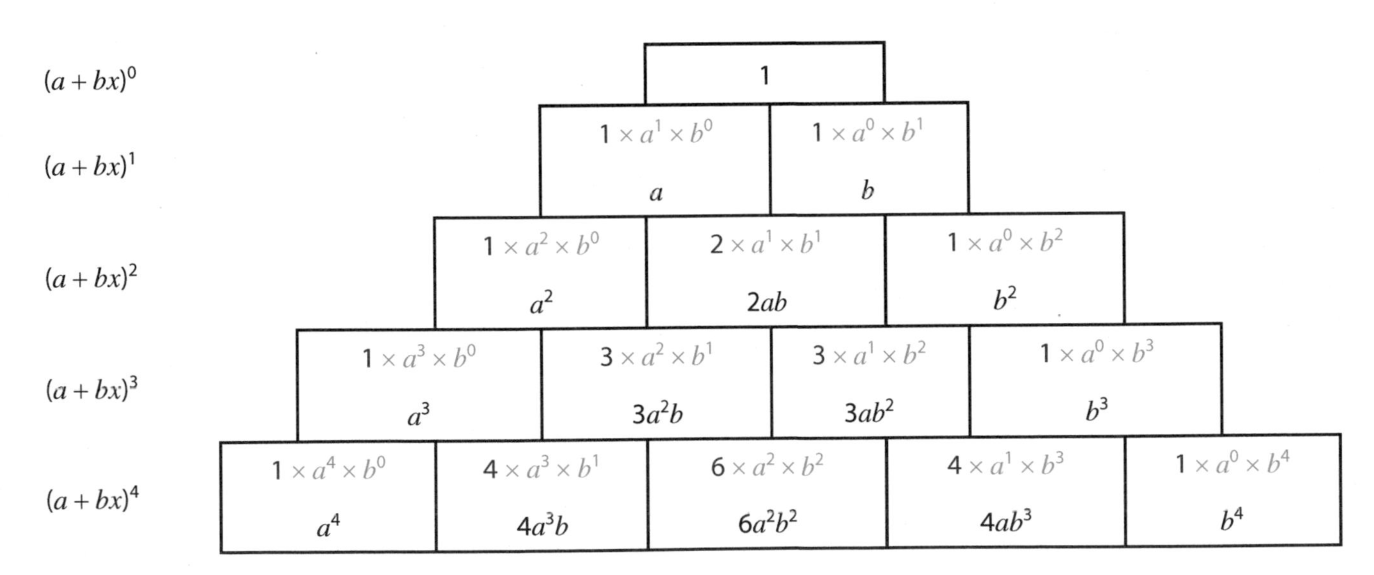

DAY 2

You can save time if you know that the first two coefficients of any expansion are always 1 and n; the symmetry of Pascal's triangle also means, for example, that ${}^7C_2 = {}^7C_5 = 21$.

Example

Expand $(2+3x)^5$ fully.

$$(2+3x)^5 = {}^5C_5 \times 2^5 + {}^5C_4 \times 2^4 \times (3x)^1 + {}^5C_3 \times 2^3 \times (3x)^2 + {}^5C_2 \times 2^2 \times (3x)^3 + {}^5C_1 \times 2^1 \times (3x)^4 + {}^5C_0 \times 2^0 \times (3x)^5$$

$$= 32 + 240x + 720x^2 + 1080x^3 + 810x^4 + 243x^5$$

Questions often ask for an expansion up to a certain term, rather than expecting the full expansion. This means larger powers of n can be used without lots of time being spent crunching numbers. The introduction of algebraic coefficients makes the topic slightly harder. Watch out for negatives as well!

Example

Expand $(3p - x)^{12}$ up to the x^2 term.

$$(3p - x)^{12} = {}^{12}C_{12} \times (3p)^{12} + {}^{12}C_{11} \times (3p)^{11} \times (-x) + {}^{12}C_{10} \times (3p)^{10} \times (-x)^2 + \cdots$$

$$= 531\,441p^{12} - 2\,125\,764p^{11}x + 3\,897\,234p^{10}x^2 + \cdots$$

Equating coefficients allows the formation of equations that can be used to solve problems.

Example

The coefficient of the x^3 term in the expansion of $(a-3x)^8$ is −48 384. Find the value of a.

$${}^8C_3 \times (-3)^3 \times (a)^5 = -48384$$

Using brackets around both terms means the calculator will correctly deal with any negatives.

$$56 \times -27 \times a^5 = -48384$$

$$a^5 = \frac{-48384}{-1512} = 32$$

$$a = \sqrt[5]{32} = 2$$

Using Binomial Expansions to Estimate a Value

You can use binomial expansion to estimate values.

Example

Use the binomial expansion of $(2-3x)^3$ to find the value of 1.7^3 to 3 decimal places. You must justify your answer.

$(2-3x)^3 = 8 - 36x^1 + 54x^2 - 27x^3$

$2 - 0.3 = 1.7$, so let $x = 0.1$

$1.7^3 = 8 - 3.6 + 0.54 - 0.027 = 4.913$

Justifying your answer means writing 4.913 is not enough to gain the mark since this could be done just using a calculator. The question asks you to use binomial expansion, so you need to show this step.

Links to Other Concepts

- Creating an expression that can then be differentiated or integrated
- Probability, permutations and combinations
- Manipulating and simplifying algebraic expressions
- Simplifying numbers that involve a surd

SUMMARY

- **A binomial of the form $(1 + x)^n$ will have coefficients for its terms which match the numbers in the corresponding row from Pascal's triangle.**
- **These values can be calculated using ^nC_r.**
- $(a+b)^n = a^n + {}^nC_1a^{n-1}b^1 + {}^nC_2a^{n-2}b^2 + \cdots + {}^nC_ra^{n-r}b^r + \cdots + b^n$

  where $^nC_r = \binom{n}{r} = \frac{n!}{r!(n-r)!}$
- **In the expansion of $(a + b)^n$, each term is $^nC_r \times a^r \times b^{n-r}$. Note the power of a and the power of b add up to n.**
- **In $(a + b)^n$ the letters a and b can represent constants but also simple functions. If the function affects the power of x, it is easy to make mistakes. Showing working and taking time helps to reduce mistakes.**
- **Coefficients can be huge or tiny, rational or irrational. Think about whether the size of any answer makes sense with the numbers involved in the question.**
- **Interpretation and application of the laws of indices and surd manipulation may be required.**

QUICK TEST

1. Expand the brackets $(3x-2)^3$.
2. Find the value of $^{16}C_5$.
3. Find the coefficient of the x^{11} term in the expansion of $(1-2x)^{11}$.
4. Find the coefficients of the x^5 and x^6 terms in the expansion of $(x-1)^{11}$.
5. Use binomial expansion to express $(2-\sqrt{3})^3$ in the form $a+b\sqrt{3}$.
6. What is the coefficient of the y^3 term in the expansion of $\left(\frac{1}{2}+4y\right)^5$?
7. What is the coefficient of the y^7 term in the expansion of $\left(a-\frac{y}{2}\right)^{12}$?
8. The coefficient of the x^2 term in the expansion of $(a+x^2)^3$ is 12. Find the value of a.

PRACTICE QUESTIONS

1. Find the first three terms, in decreasing powers of x, of the binomial expansion of $\left(3x+\frac{2}{x}\right)^5$. **[3 marks]**

2. A binomial expansion gives the result $(a+f(x))^n = 2401 - 2744x + 1176x^2 + px^3 + qx^4$, where a, p and q are constant integers and $f(x)$ is a simple function of x.

a) i) Given that $a < 5$, what is the value of a? **[1 mark]**

ii) Find $f(x)$. **[2 marks]**

b) Find the values of p and q. **[2 marks]**

c) Hence or otherwise find $\frac{d}{dx}(2a+f(x))^4$. **[4 marks]**

3. The expression $10\left(2x-\frac{1}{2}\right)^4-(a-x)^5$ has no x^3 term. Find the positive integer a. **[4 marks]**

4. a) Express $^nC_2$ as a fraction in its simplest form, without factorials. **[1 mark]**

The binomial expansion of $(3x+1)^n$, where n is positive, has an x^2 coefficient of 252.

b) Find n. **[4 marks]**

Trigonometry 1

Trigonometry uses the relationship between angles in triangles and the ratio of the sides. Questions will be in degrees at AS-level and calculators should be set up accordingly. The basic trigonometric ratios are:

$\sin\theta = \frac{\text{opposite}}{\text{hypotenuse}}$ $\cos\theta = \frac{\text{adjacent}}{\text{hypotenuse}}$ $\tan\theta = \frac{\text{opposite}}{\text{adjacent}}$

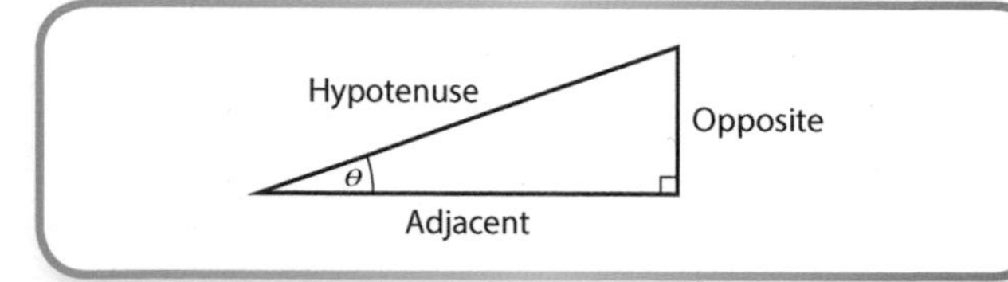

Non-Right-Angled Triangles

The following trigonometric equations can be used with non-right-angled triangles.

Area of Triangle $= \frac{1}{2}ab\sin C$

This equation is used for any question that involves area. It is set up to find the **area** given two sides and the angle between them. It can be rearranged to find a side or angle if the area is given.

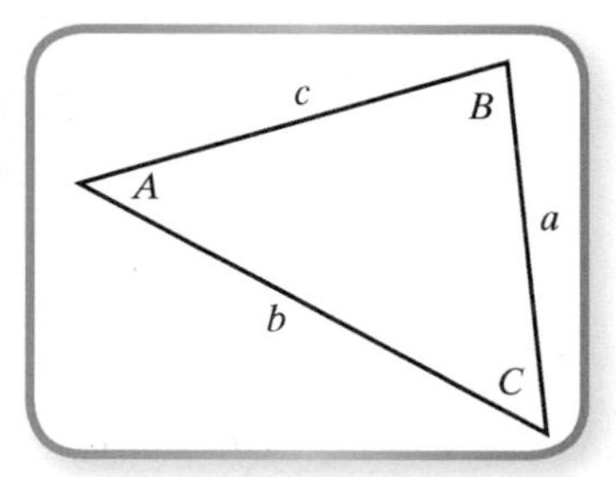

Example
Find the area of this triangle correct to 3 significant figures.

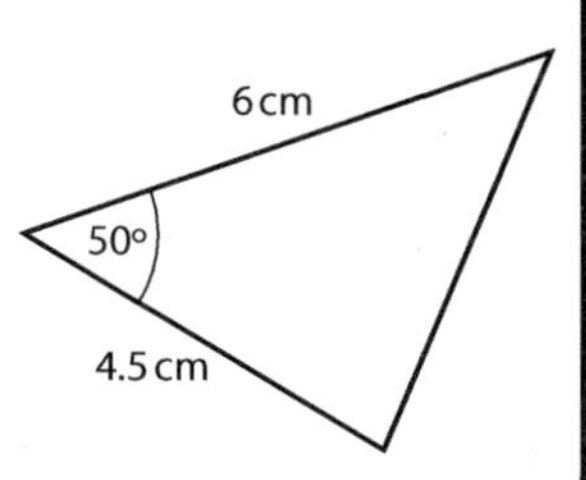

Area $= \frac{1}{2} \times 4.5 \times 6 \times \sin 50$

$= 10.3\ \text{cm}^2$ (3 s.f.)

Example
Given that the area of triangle ABC is 12 and sides AB = 4 and AC = 7, find angle A to 1 decimal place.

$12 = \frac{1}{2} \times 4 \times 7 \times \sin A$

$\sin A = \frac{12}{\frac{1}{2} \times 4 \times 7} = \frac{6}{7}$

$A = \sin^{-1}\left(\frac{6}{7}\right) = 58.9972\ldots = 59.0°$ (1 d.p.)

The Sine Rule: $\frac{a}{\sin A} = \frac{b}{\sin B} = \frac{c}{\sin C}$

The sine rule can also be written $\frac{\sin A}{a} = \frac{\sin B}{b} = \frac{\sin C}{c}$. This equation is used when given an angle–side pair and then either an angle or a side and you are asked to find the **corresponding** angle or side (i.e. when two angles and two sides are involved, one of which is the value to be found).

Use with **internal angles** of a triangle (which add up to 180°) to create the required angle–side pair if needed.

The ambiguous case: It is possible if given angle A, side a and side c (i.e. SSA, two sides and an angle that isn't between them) that there are two possible positions for the line BC, as shown in the diagram.

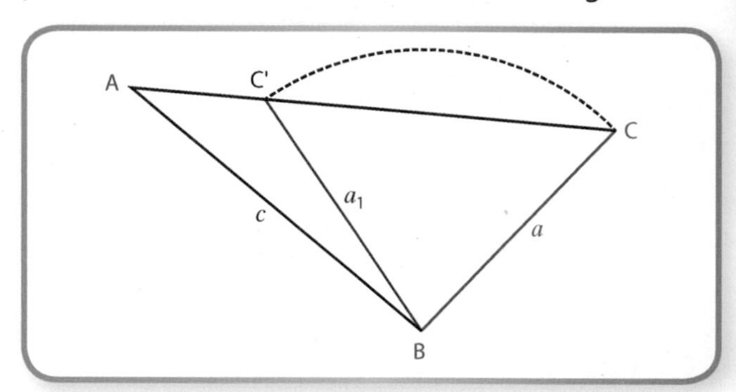

The question may hint at which is the correct one but it is important to acknowledge both then reject one with a reason, or give both as a final answer if they are relevant.

By recognising an **SSA** case and finding two possible answers, you can check if both are valid by making sure that the sum of the obtuse angle found and the original angle given is less than 180°. Calculators will only give one result, θ. To find the second result, use $180 - \theta$. You should expect two answers to be valid if given S_1S_2A, where S_1 is opposite angle A, such that $S_1 < S_2$.

Example

Triangle PQR has sides PR = 11 cm, RQ = 6 cm and angle $P = 27°$.

a) Find the size of angle Q, giving your answer to 1 decimal place.

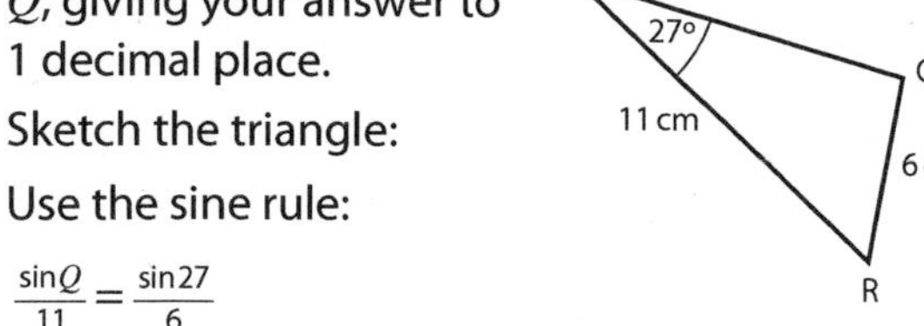

Sketch the triangle:

Use the sine rule:

$$\frac{\sin Q}{11} = \frac{\sin 27}{6}$$

$$\sin Q = \frac{11 \sin 27}{6}$$

$Q = 56.3373\ldots$ (store in calculator as A to retain accuracy)

As SSA case, is there a second result?

$180 - 56.337\ldots = 123.66\ldots$ (store in calculator as B)

As $27 + 123.66\ldots < 180$ $\therefore$ both results are valid.

$Q = 56.3°$ or $123.7°$ (1 d.p.)

b) The perimeter of the shape is less than 28 cm. Use the sine rule to find the size(s) of side PQ.

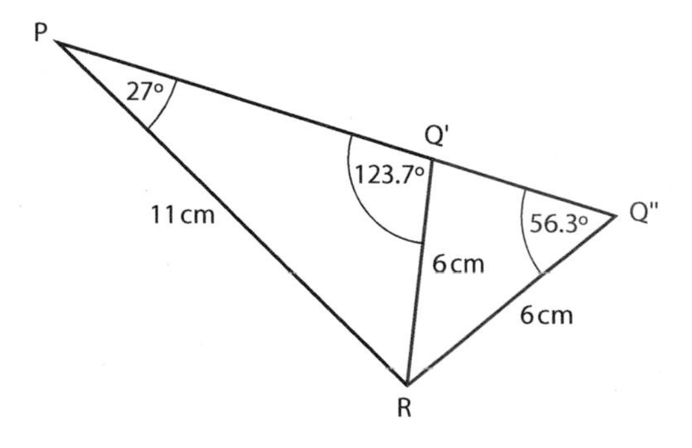

The two options for the position of Q are shown.

If perimeter < 28 cm, then PQ < 28 – (11 + 6) = 11 cm

Using triangle PRQ′:

$$\frac{11}{\sin 123.7} = \frac{PQ'}{\sin(180 - 27 - 123.7)}$$

$$\Rightarrow PQ' = \frac{11}{\sin 123.7} \times \sin 29.3 = 6.47$$

Using triangle PRQ″:

Q′R = Q″R = 6 cm, so triangle RQ′Q″ is isosceles and

Q′Q″ = 2(6 cos 56.3) = 6.66 cm and

PQ″ = 6.47 + 6.66 = 13.1 cm > 11 cm

Therefore PQ = 6.47 cm

The Cosine Rule: $a^2 = b^2 + c^2 - 2bc \cos A$

The cosine rule is used to find an unknown side when given a pair of sides and the angle between them. It can also be rearranged to find b, c or A. Finding b or c will involve solving a quadratic. The cosine rule is used if only one angle is involved, whether it is the unknown or a given, i.e. when three sides and an angle are involved in the question, one of which is the unknown.

Example

A triangle ABC has an area of 15 cm². Side AC = 5 cm and angle $C = 60°$. Find the length of line AB, correct to 3 significant figures.

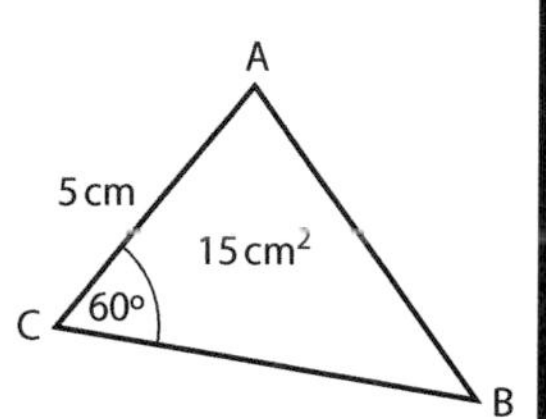

Start by finding a side using the area rule:

$$15 = \frac{1}{2} \times 5 \times BC \times \sin 60$$

$$BC = \frac{15}{\frac{1}{2} \times 5 \times \sin 60} = 4\sqrt{3}\text{ cm}$$

To find AB, now apply the cosine rule (as there is only one angle): $a^2 = b^2 + c^2 - 2bc\cos A$

$$AB^2 = (4\sqrt{3})^2 + 5^2 - 2 \times 5 \times 4\sqrt{3} \times \cos 60$$

$$= 73 - 20\sqrt{3}$$

$$AB = \sqrt{73 - 20\sqrt{3}} = 6.1934\ldots = 6.19\text{ cm (3 s.f.)}$$

Don't forget

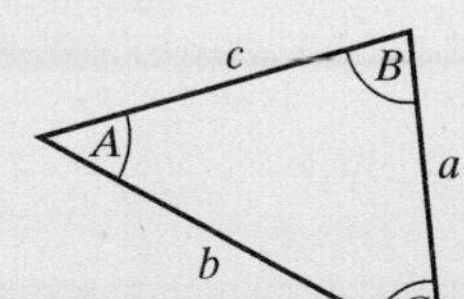

Sketching the triangle is always a good idea!

Conventions related to labelling triangles mean that the side BC is the side opposite the given angle A. This means in a general triangle ABC:

a = BC, b = AC, c = AB

The a, b and c in the equation may not match the ABC in the set-up of the triangle. Be careful! The key thing with the cosine rule is that the side 'a' is opposite the angle in question. In the example above, the angle is at point C so either the triangle could be relabelled such that A is the known angle or the equation could be treated as $c^2 = a^2 + b^2 - 2ab \cos C$.

DAY 2

Graphs of Sine, Cosine and Tangent

Whilst the roots and proofs of trigonometric ratios lie in triangles, it is important to remember that sine, cosine and tangent are many-to-one functions. This means it is possible to look at angles beyond the 180° in a triangle. For any given ratio, there are many possible angles that would give the required result. Each x-value gives only one y-value, though.

The key features of the sine, cosine and tangent graphs are shown below.

$y = \sin x$

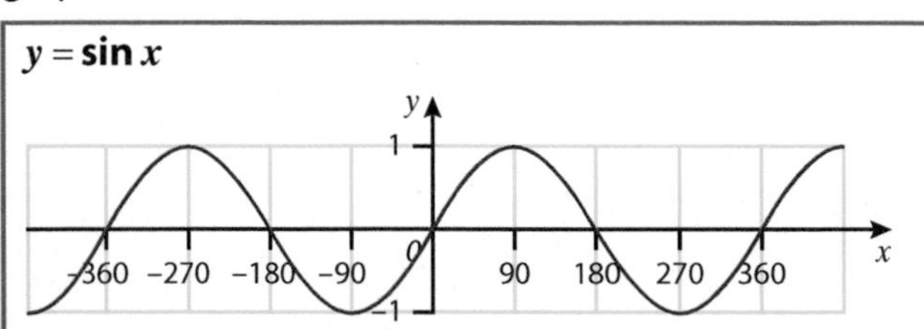

- Maximum value 1, minimum value −1 and y-intercept = 0
- Vertical lines of symmetry through every minimum and maximum. …, −270°, −90°, 90°, 270°, …
- Rotational symmetry about the origin $\rightarrow \sin(\alpha) = -\sin(-\alpha)$
- Periodicity 360° $\rightarrow \sin(\beta) = \sin(\beta + 360n)$, where n is an integer value

$y = \cos x$

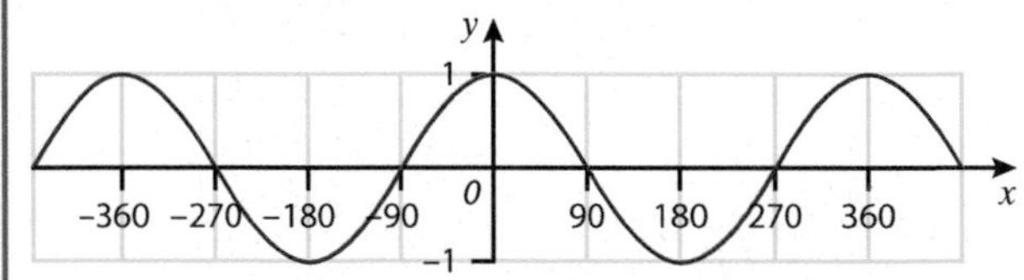

- Maximum value 1, minimum value −1 and y-intercept = 1
- Vertical lines of symmetry through every minimum and maximum. …, −360°, −180°, 0°, 180°, 360°, …
- Reflectional symmetry about the y-axis $\rightarrow \cos(\alpha) = \cos(-\alpha)$
- Periodicity 360° $\rightarrow \cos(\beta) = \cos(\beta + 360n)$, where n is an integer value

$y = \tan x$

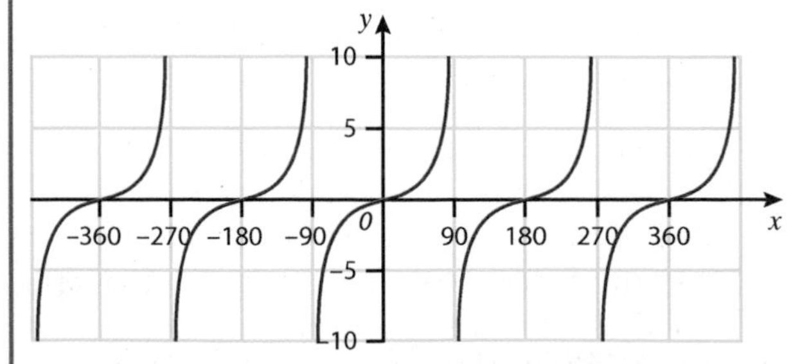

- Vertical asymptotes at …, −270°, −90°, 90°, 270°, …
- Periodicity 180° $\rightarrow \tan(\beta) = \tan(\beta + 180n)$, where n is an integer value
- y-intercept = 0
- Between each pair of asymptotes, the graph has rotational symmetry about the x-intercept
- No lines of symmetry

When solving a trigonometric equation, using these properties of the graphs allows you to find all the possible solutions within a given range. Sketching the relevant graph can help to identify the number of solutions and how to find them.

Example

Find all the solutions to $\sin^2(x) = 0.25$ in the range $0° \leqslant x \leqslant 360°$.

$\sin(x) = \pm\sqrt{0.25} = \pm 0.5$

$\sin^{-1}(0.5) = 30°$

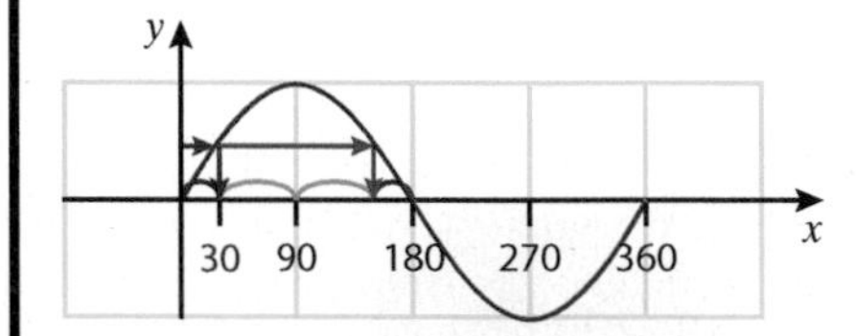

Calculator result 30°

Based on symmetry about $x = 90$, there is an additional result at $\boxed{180 - 30} \rightarrow 150 \leftarrow \boxed{90 + 60}$

$\sin^{-1}(-0.5) = -30°$, which is outside the required range but can be used to find the solutions within the range.

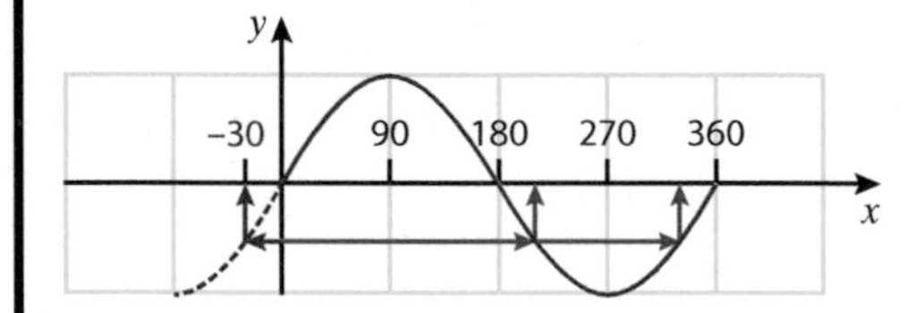

Calculator result −30°

Additional results of: $180 + 30 = 210°$
$-30 + 360 = 330°$

$x = 30°, 150°, 210°, 330°$

Note: Substituting results back into a calculator is a way of double checking them, e.g. $\sin(210) = -0.5$.

Links to Other Concepts

- Bearings, triangles, coordinate geometry
- Transformations of graphs
- Circles and lines
- Proof
- Pythagoras' Theorem
- Vectors and mechanics
- Surd manipulation, rationalising denominators

SUMMARY

- **Sine, cosine and tangent are the three key trigonometric functions used at AS-level.**
- **Area of triangle $= \frac{1}{2}ab\sin C$**
- **The sine rule: $\frac{a}{\sin A} = \frac{b}{\sin B} = \frac{c}{\sin C}$**
- **Watch out for SSA cases when applying the sine rule – check for two possible results.**
- **The cosine rule: $a^2 = b^2 + c^2 - 2bc\cos A$**
- **θ, α and β are often used to denote angles.**
- **If dividing through by a trigonometric function which is a common factor in all terms in the equation, that function equalling zero provides a valid result that can be easily forgotten.**

QUICK TEST

1. Find the value of tan(83).
2. Find the unknown side in these:

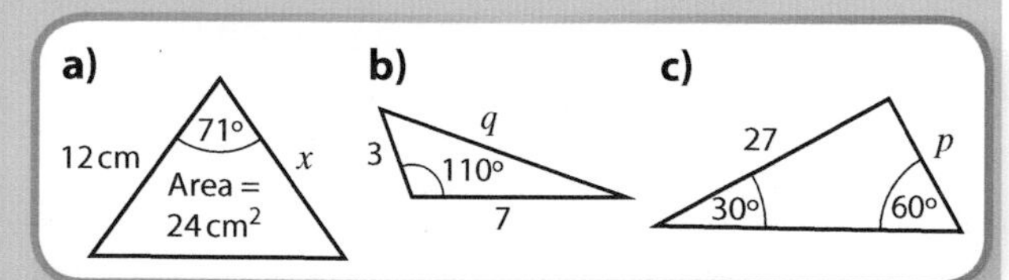

3. Find the value of angle θ in these:

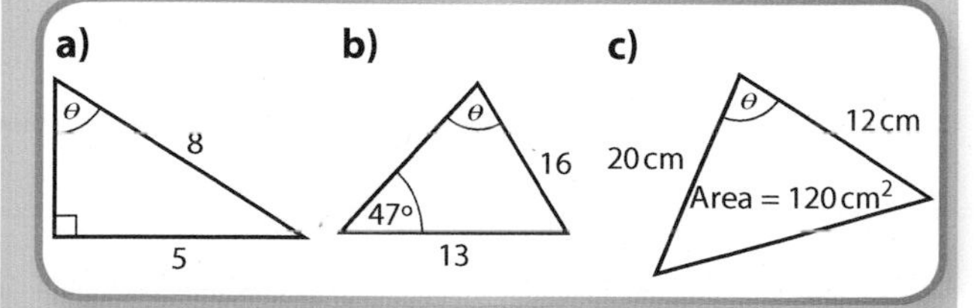

4. How many solutions are there to the equation $3\sin^2(x) + 2\sin x = 0$ in the range $0 \leqslant x < 720°$?
5. Calculate the exact value of $14\sin(75)(-\cos(-270) - 2\tan(135))$.

PRACTICE QUESTIONS

1. $4\cos^2\alpha + 10\cos\alpha - 6 = 0$

   a) Find the value of $\cos\alpha$. **[2 marks]**

   b) Hence find all the values of α in the range $-180° \leqslant \alpha \leqslant 360°$. **[2 marks]**

   c) Hence or otherwise find all the possible solutions to $2\cos^3\alpha + 5\cos^2\alpha = 3\cos\alpha$. **[3 marks]**

2. $(3\cos x - 1)(5\sin x - 2)\tan x = 0$. Find all the possible values of x in the range $-180° \leqslant x \leqslant 360°$. **[4 marks]**

3. Kat and Rob are cycling on a forest path, which splits into two straight paths at an angle of 30° to each other. Kat, who cycles twice as fast as Rob, takes the left path and Rob takes the right path.

   a) Write down an expression in terms of x for the **exact** distance between Kat and Rob as they cycle along the paths, where x is the distance cycled by Rob from the split. **[4 marks]**

   As the path Rob is cycling along emerges from the forest, he looks to his left and can see Kat.

   b) What angle would Kat need to look back through in order to be looking directly at Rob at this point? Give your answer to an appropriate degree of accuracy. **[4 marks]**

   c) Give a reason why the model isn't realistic. **[1 mark]**

Trigonometry 2

The previous topic looked at solving equations with a single trigonometric function involved. More complex equations might require the use of trigonometric identities to simplify an equation to the point where it can be solved using the methods met previously.

$$\frac{\sin\theta}{\cos\theta} = \frac{\frac{\text{opp}}{\text{hyp}}}{\frac{\text{adj}}{\text{hyp}}} = \frac{\text{opp}}{\text{adj}} \rightarrow \mathbf{\tan\theta \equiv \frac{\sin\theta}{\cos\theta}}$$

Also note that $\tan^2\theta \equiv (\tan\theta)^2 \equiv \left(\frac{\sin\theta}{\cos\theta}\right)^2 \equiv \frac{\sin^2\theta}{\cos^2\theta}$

Example

$4\cos\theta\tan\theta = 1$. Find all the solutions to θ in the range $0° < \theta \leqslant 360°$ accurate to 1 decimal place.

Substitute in $\tan\theta \equiv \frac{\sin\theta}{\cos\theta}$

$4\cos\theta\frac{\sin\theta}{\cos\theta} = 1$

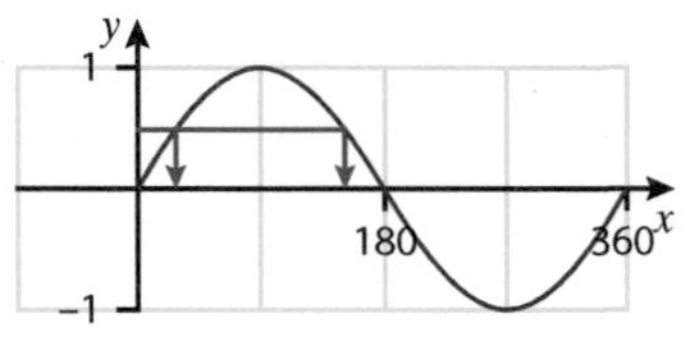

$4\sin\theta = 1$

$\sin\theta = \frac{1}{4}$

$\theta = 14.4775\ldots = 14.5°$(1 d.p.)

$\theta = 180 - 14.4775\ldots = 165.522\ldots° = 165.5°$ (1 d.p.)

$\sin^2\theta + \cos^2\theta \equiv 1$

$\sin^2\theta + \cos^2\theta \equiv 1$ can be rearranged to give:

$\sin^2\theta \equiv 1 - \cos^2\theta$ and **$\cos^2\theta \equiv 1 - \sin^2\theta$**

This might be used to simplify an equation or to remove a function from the equation so that it can be solved. It could also be used to find the corresponding sine and cosine values given one or the other, in their exact form.

Example

Given that $\sin\theta = \frac{2}{5}$, and θ is an acute angle:

a) Find the exact value of $\cos\theta$.

By finding θ, then substituting into cos, the answer from a calculator would be 0.916 51…. Another method is needed to find this as an exact value.

Using the identity $\cos^2\theta \equiv 1 - \sin^2\theta$

$\cos^2\theta = 1 - \left(\frac{2}{5}\right)^2$

$\cos^2\theta = \frac{21}{25}$

$\cos\theta = \sqrt{\frac{21}{25}} = \frac{\sqrt{21}}{5}$

b) Hence or otherwise find the value of $\tan\theta$.

$\tan\theta = \frac{\sin\theta}{\cos\theta} = \frac{\frac{2}{5}}{\frac{\sqrt{21}}{5}}$

$\tan\theta = \frac{2}{\sqrt{21}} = \frac{2\sqrt{21}}{21}$

As the question states that θ is acute, both $\tan\theta$ and $\cos\theta$ will be positive.

c) If θ was obtuse instead of acute, what would the value of $\cos\theta$ be?

Looking at the graphs of sine and cosine together, you can see that the value of $\cos\theta$ will now be negative but of the same magnitude. (Consider rotational symmetry.)

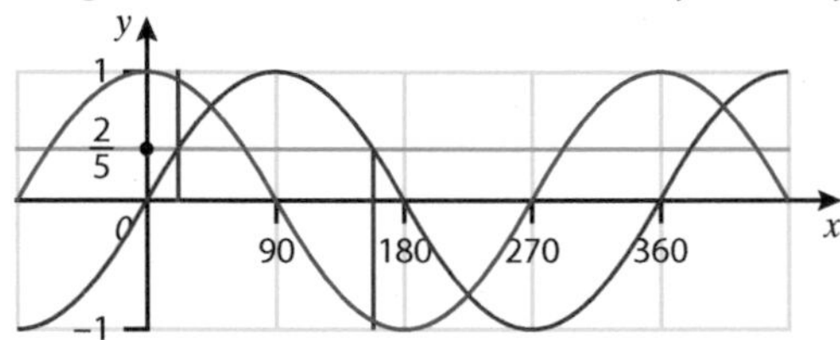

$\cos\theta = -\frac{\sqrt{21}}{5}$

Using the CAST diagram would give the same result and is some people's preferred method:

Acute – all positive

$\cos\theta = \frac{\sqrt{21}}{5}$

$\tan\theta = \frac{2}{\sqrt{21}} = \frac{2\sqrt{21}}{21}$

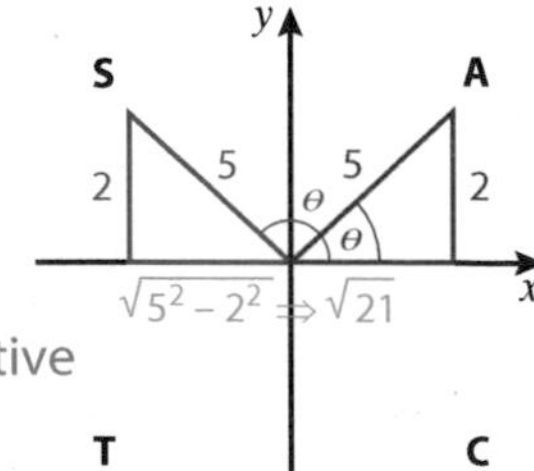

Obtuse – only sine positive

$\cos\theta = -\frac{\sqrt{21}}{5}$

$\tan\theta = -\frac{2}{\sqrt{21}} = -\frac{2\sqrt{21}}{21}$

Look out for when the equation forms a quadratic. Remember both positive and negative values when taking a square root. Sometimes both answers will be valid. At other times, an answer can be discounted as it is too large or too small (e.g. $\cos\theta = 2$, since cosine has a maximum value of 1 and a minimum of -1).

Often it is hard to know where to start with trigonometric identity questions. Some ways may be more efficient; others might take you round in a circle. Don't spend too long deliberating; get stuck in and see what happens.

Key things to look out for:

- Mixed trigonometric functions – is it possible to get the equation in terms of just one?
- Trigonometric functions in denominators – if it makes it look more complicated, consider multiplying through by the denominator to see what happens.
- What is the bit of the equation that seems most complicated and is there a way to simplify it (multiplying out brackets, multiplying up by denominators, etc.)?

Example

Given that $3\tan\theta = 2\cos\theta$:

a) Show that $2\sin^2\theta + 3\sin\theta - 2 = 0$.

Both tangent and cosine are present (and not in separate brackets when equal to 0). The final target given shows that the aim is to get it in terms of sine.

$3\tan\theta = 2\cos\theta$

Always start from the original statement and show small steps.

$3\frac{\sin\theta}{\cos\theta} = 2\cos\theta$

Substituting in for tangent, using the identity, still leaves cosine and sine:

$3\sin\theta = 2\cos^2\theta$

Multiplying up by cosine creates cosine-squared, for which we have another identity:

$3\sin\theta = 2(1 - \sin^2\theta)$

Once substituted it is a relatively simple algebraic rearrangement:

$3\sin\theta = 2 - 2\sin^2\theta$

$2\sin^2\theta + 3\sin\theta - 2 = 0$

b) Hence or otherwise find the solutions for θ in the range $-180° < \theta \leqslant 180°$.

$2\sin^2\theta + 3\sin\theta - 2 = 0$

Recognising quadratics is important.

$(2\sin\theta - 1)(\sin\theta + 2) = 0$

Factorisation is great for simple quadratics but completing the square or the quadratic formula are just as good.

$\sin\theta + 2 = 0 \Rightarrow \sin\theta = -2$, which is impossible

Acknowledging a result is important, even if you disregard it. It shows you haven't forgotten about it.

$2\sin\theta - 1 = 0 \Rightarrow \sin\theta = \frac{1}{2}$

$\sin^{-1}\left(\frac{1}{2}\right) = 30°$

$\theta = 30°, 150°$

Multiples of the Unknown Angle

Using your knowledge of transformations of graphs, you can solve problems where the angle within the function has a multiplier.

$y = \mathbf{f}(ax)$ gives a graph that is stretched horizontally by a factor of $\frac{1}{a}$.

There are different ways of approaching this type of question. Two key methods are detailed below.

Method 1: Extending the Range

Solve for $a\theta$ and find a set of results. The range stated in the question has to be adjusted so all the results for θ are found.

Method 2: Transforming the Graph

Sketch the transformed graph, then use its symmetry and periodicity to solve for θ.

Example

Find all the solutions of $\cos 2x = 0.6$ in the range $-180° < x \leqslant 180°$.

Method 1

The range needs to be doubled to find all the values that, when halved, would fall into the range $-180° < x \leqslant 180°$.

Let $\alpha = 2x$

Solve $\cos\alpha = 0.6$ in the range $-360° < \alpha \leqslant 360°$.

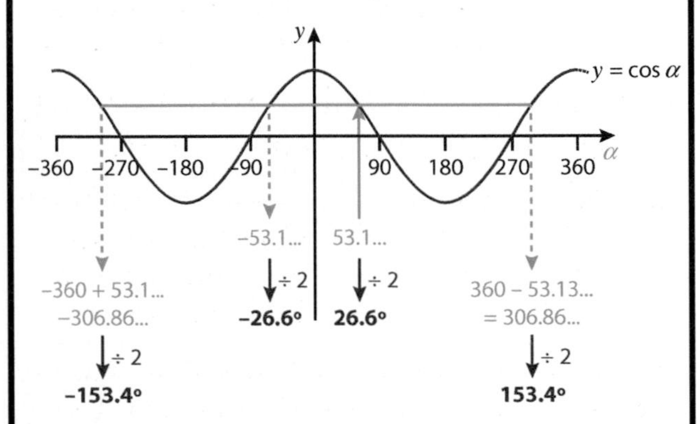

Remember to halve for the final answer.

Method 2

The graph is squashed by half. Having found one result for x, drawing the transformed graph allows the rest of the results to be found.

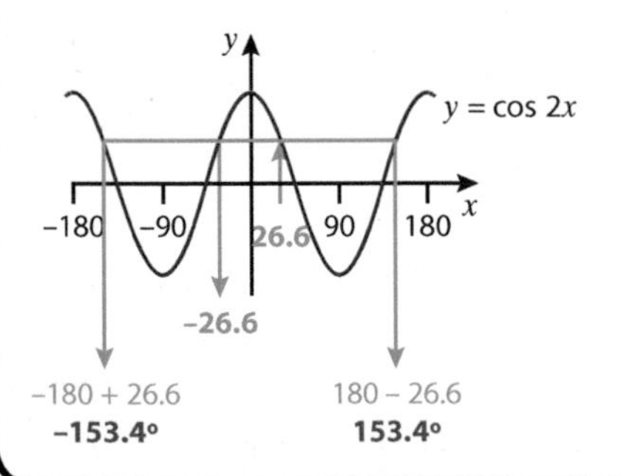

Pulling It All Together

Questions will generally include more than one element. As with algebraic manipulations, look for a starting point and see what happens. Generally, the aim is to recognise how the equation can be simplified as much as possible. Often, part a) of a question will be a 'show that', meaning part b) can be attempted even if part a) wasn't successfully achieved.

Example

Given that $1 + \tan^2 3\theta = \frac{1}{\cos^2 3\theta}(7 - \tan 3\theta)$, find all the solutions for θ in the range $-180° < \theta \leqslant 180°$.

Let $\alpha = 3\theta$ and solve in the range $-540° < \alpha \leqslant 540°$.

$1 + \tan^2\alpha = \frac{1}{\cos^2\alpha}(7 - \tan\alpha)$

$\cos^2\alpha + \cos^2\alpha\tan^2\alpha = 7 - \tan\alpha$

$\cos^2\alpha + \cos^2\alpha\frac{\sin^2\alpha}{\cos^2\alpha} = 7 - \tan\alpha$

$\cos^2\alpha + \sin^2\alpha = 7 - \tan\alpha$

$1 = 7 - \tan\alpha \quad \rightarrow \quad \tan\alpha = 6$

$\alpha = \tan^{-1}(6) = 80.537677\ldots°$

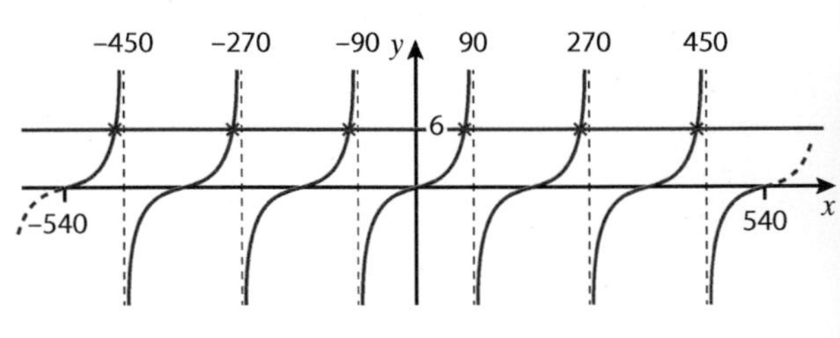

$\alpha = -459.4\ldots°, -279.4\ldots°, -99.4\ldots°, 80.5\ldots°, 260.5\ldots°, 440.5\ldots°$

$\alpha = 3\theta \rightarrow \theta = \frac{\alpha}{3}$

$\theta = -153.2°, -93.2°, -33.2°, 26.8°, 86.8°, 146.8°$ (answers to 1 d.p.)

Links to Other Concepts

- Surds
- Bearings, triangles, coordinate geometry
- 'Hidden' quadratics giving a positive and negative result for a trigonometric ratio
- Transformations of graphs
- Proof, justification and algebraic manipulation

SUMMARY

- **Trigonometric identities are rules that are always true for any value of θ.**
- **$\tan\theta \equiv \frac{\sin\theta}{\cos\theta}$ is used to convert between the trigonometric functions. Turn it into tangent to simplify and remove sines and cosines. Turn it into sine and cosine to cancel terms, use other identities and move towards a result.**
- **$\sin^2\theta + \cos^2\theta \equiv 1$ is used to simplify and convert between sine and cosine. It can also be used to find the exact values of the other trigonometric functions.**
- **Sketches of the functions will help to find all the results in a given range.**
- **Quadratics are common in this topic and need to have all results considered.**
- **Exact results will generally contain surds. If asked for an exact result, a decimal answer from a calculator is not acceptable, but it could be used to check the accurate answer gained from a different method.**
- **When applying identities, the angle must be exactly the same.**
- **When solving an equation where the angle has a multiplier, adjust the range or use the transformed graph to get all the results.**

QUICK TEST

1. Given that $\cos^2\alpha = 0.34$, find the value of $\sin^2\alpha$.
2. Find all the results for $3\sin x = 5\cos x$ in the range $0° < x < 360°$.
3. Given that $\sin\alpha = \frac{3}{7}$ and $\tan\alpha = \frac{-3\sqrt{10}}{20}$, find the value of $\cos\alpha$.
4. Find the lowest positive value of θ where the graph $y = \sin\theta$ meets $y = \cos\theta$.
5. How many solutions are there to the equation $\cos^2 x = \frac{1}{4}$ in the range $-360° < x < 360°$?
6. It is given that $(4 - 7\cos\theta)(5\cos\theta + 2) = 0$. Find the values of θ in the range $0° < x < 360°$.
7. Sketch the graphs of $y = \sin x$ and $y = \tan 2x$ for the range $0° \leqslant x \leqslant 360°$ on the same set of axes, showing clearly all points that represent a solution to the equation $\sin x = \tan 2x$.
8. Lou says that all the solutions to $\sin x = \tan 2x$ in the range $0° \leqslant x \leqslant 360°$ can be found by dividing by tan to give the equation $\cos x = 1$.
   a) Lou is incorrect. Explain why.
   b) Lou says that her answer at least finds all the results to $\sin x = \tan x$. Is she correct? Justify your answer.

PRACTICE QUESTIONS

1. Imran is trying to solve the equation $\sin x(12\sin x\tan x - 5) = 10\tan x$ in the range $-180° < \theta \leqslant 180°$. It can be simplified to a quadratic in terms of $\cos x$, which Imran says will find all the possible solutions.
   a) The quadratic is in the form $a\cos^2 x + b\cos x + c = 0$. Find a, b, and c. **[4 marks]**
   b) Is he correct that it will find all the possible solutions to the original equation? **[1 mark]**
   c) Find all the solutions to the equation within the given range. **[5 marks]**

2. a) Given that $k\sin x = 5$, state the values of k for which there is no solution to this equation. **[2 marks]**
   b) The equation $2k\sin^2 x + (k - 10)\sin x - 5 = 0$ factorises to $(2\sin x + a)(k\sin x + b) = 0$. Find a and b. **[2 marks]**
   c) There are three solutions to $2k\sin^2 x + (k - 10)\sin x - 5 = 0$ in the range $180° < x \leqslant 360°$. Find the value of k and the values of x. **[4 marks]**

3. Find the solutions to the equation $24\cos^2\left(\frac{\theta}{2}\right) = 2\sin\left(\frac{\theta}{2}\right) + 19$ in the range $-720° < \theta \leqslant 720°$. **[7 marks]**

Exponentials

General Exponential Functions

Exponential comes from the word 'exponent', which in mathematics means the power or **index**. It is when f(x) is of the form $f(x) = a^x$, $a > 0$. It is defined for positive values of a only, as a **negative** number raised to a power creates a **discontinuous** set of results, the detail of which is not needed for A-level maths.

If $0 < a < 1$, the function can be rewritten with a negative index (power/exponent).

> **Example**
>
> The function $f(x) = \left(\frac{1}{3}\right)^x$ can be rewritten as $f(x) = 3^{-x}$.

Considering $a > 1$

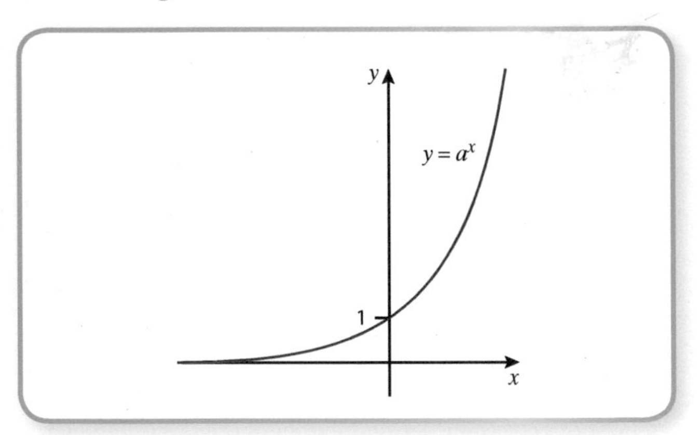

The resultant graph is a continuous curve. As x tends to negative infinity, then y tends to 0; as x tends to positive infinity, then y also tends to positive infinity. As x increases, the gradient of the curve also increases. All graphs in the form $y = a^{kx}$ will have a y-intercept of 1, as anything raised to the power of 0 is 1.

By changing the value of a, the gradient of the graph changes. A larger value of a will give a steeper curve.

Example

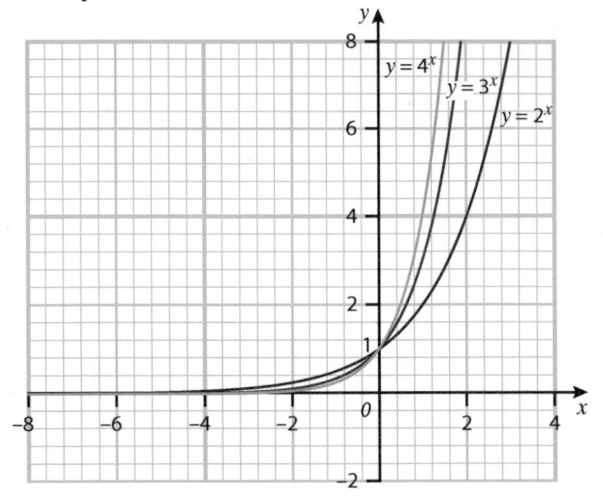

Considering $0 < a < 1$: The graphs are effectively flipped in the y-axis. As $x \to \infty$, $y \to 0$. As $x \to -\infty$, $y \to \infty$.

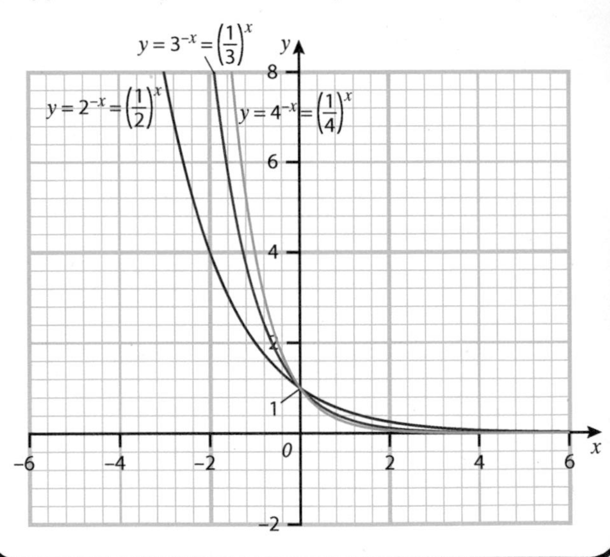

Transformations with the General Graph

Consider the graph $\mathbf{f}(x) = p^x$.

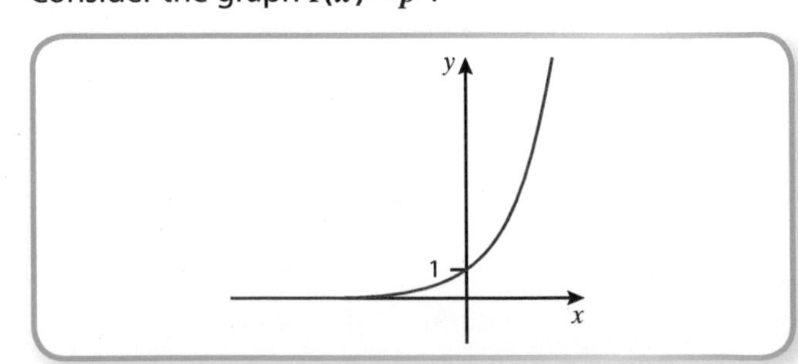

$f(x)+a=p^x+a$

The graph is moved up by a, or translated by $\begin{pmatrix}0\\a\end{pmatrix}$. If a is negative, the movement will be downwards.

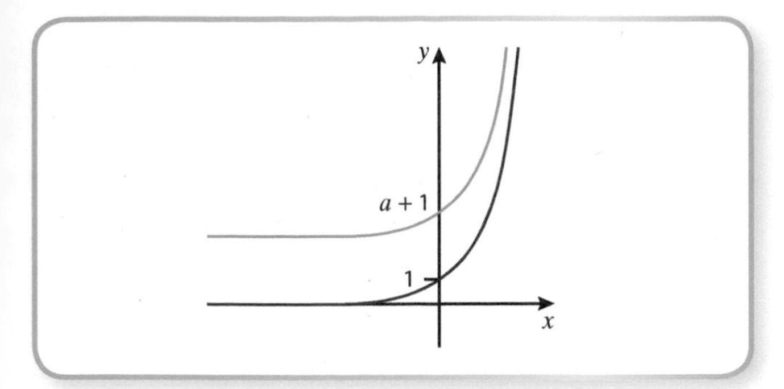

$f(x+a)=p^{x+a}$

The graph is moved left by a, or translated by $\begin{pmatrix}-a\\0\end{pmatrix}$. If a is negative, the movement will be to the right.

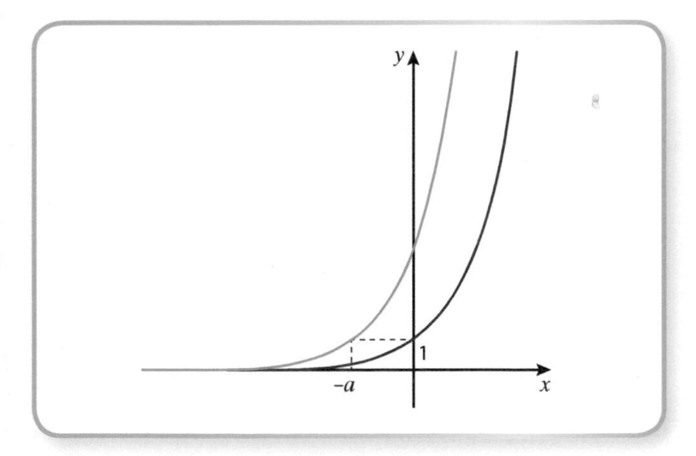

$af(x)=ap^x$, where a is a positive constant

The graph is stretched vertically by a. All x-intercepts stay in the same place. Each y-coordinate is $\times\, a$.

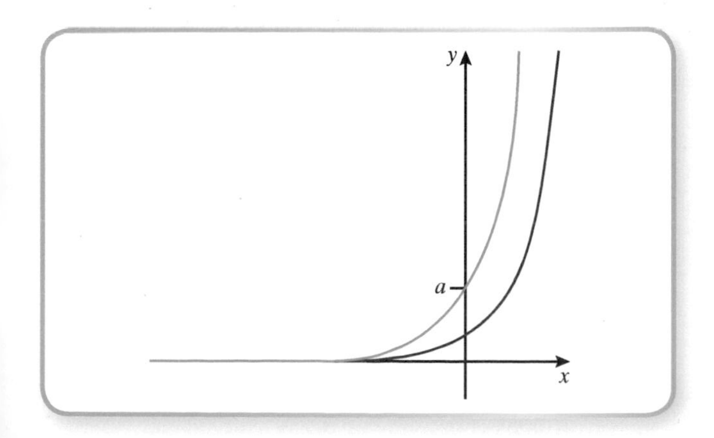

$f(ax)=p^{ax}$, where a is a positive constant

The graph is stretched horizontally by $\frac{1}{a}$. All y-intercepts stay in the same place. Each x-coordinate is $\times\,\frac{1}{a}$.

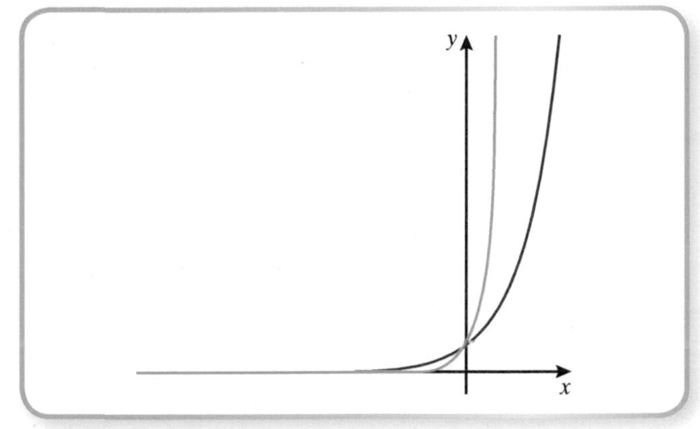

$-f(x)=-p^x$

The graph is reflected in the x-axis.

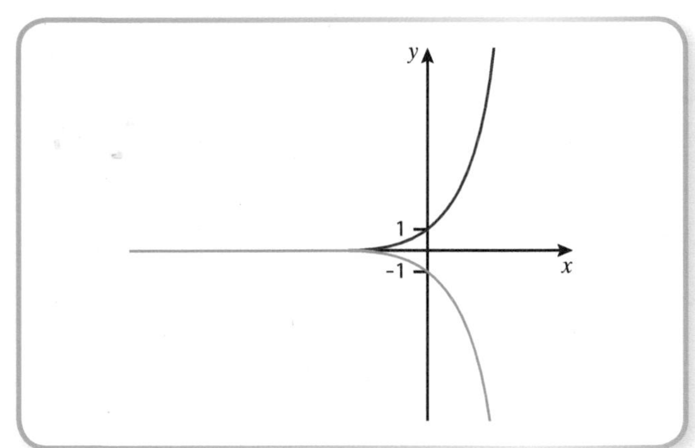

$f(-x)=p^{-x}=\left(\frac{1}{p}\right)^x$

The graph is reflected in the y-axis.

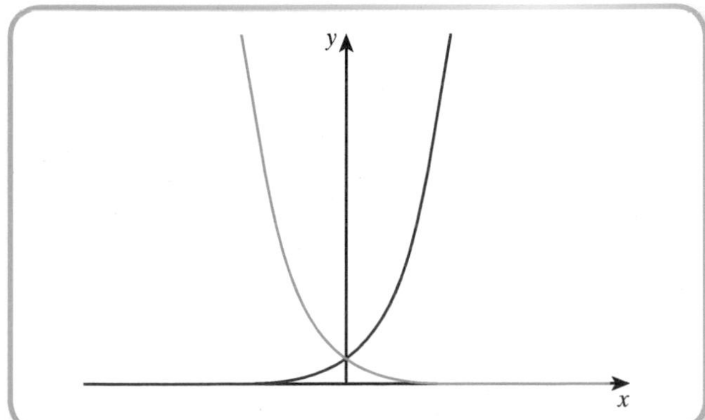

DAY 3

The exponential graphs are interesting as horizontal translations and vertical stretches are versions of each other. This is because a horizontal translation of $\begin{pmatrix} -b \\ 0 \end{pmatrix}$ is created by $f(x+b) \rightarrow a^{x+b} = a^b \times a^x$. This is the same as $a^b f(x)$, which is a vertical stretch by a factor of a^b.

Example

The function $f(x) = 3^x$ has a translation applied to it of $\begin{pmatrix} 3 \\ 0 \end{pmatrix}$.

What will be the value of the y-intercept for the translated graph? Describe the transformation as a stretch.

$f(x-3) = 3^{x-3} = 3^{-3} \times 3^x = \frac{1}{27} f(x)$

The new y-intercept will be $3^{-3} = \frac{1}{27}$. This is seen either from it being the stretch factor in the y-direction, or by substituting $x = 0$ into $f(x-3)$.

The transformation could be a stretch in the y-direction with a scale factor of $\frac{1}{27}$.

The Exponential Functions $y = e^x$

e (2.71828182…) is an important mathematical constant. Like π it is irrational, so a letter or symbol is used to represent the fully accurate number. e, like π, occurs in a number of places mathematically. In this context, e occurs as it is the value for which the gradient of the function is equal to the function. There is an e■ button on all scientific and graphical calculators.

Example

The gradient of the graph $y = e^x$ at $x = 2$ is $e^2 = 7.389056\ldots$

Sometimes the question will specify a form for the final answer. At other times the final form will be decided based on common sense and context. The important thing is to keep values accurate until the final answer.

More generally: **the gradient of the graph $y = ae^{kx}$ at any point is ake^{kx}.**

Exponential growth is used to model various real-life problems, from the spread of illness to the half-life of radioactive material and compound interest. It is used in situations where the rate or growth is proportional to the size of the population in question. The exponential function $y = e^x$ is a key part in the modelling of many situations involving growth and decay.

Example

Bacteria is grown in a culture and the rate of growth is proportional to the number of cells present. The relationship between the number of cells (y) present at any time t (in hours) after midday is $y = Ae^{kt}$. At midday there are 5000 bacteria introduced into the culture. After 2 hours there are 160 000. How many bacteria will be present in the culture at midnight? Give your answer to 4 significant figures.

First find the value of the constants A and k.

At $t = 0$

$5000 = Ae^{k \times 0}$

$A = 5000$

At $t = 2$

$160\,000 = 5000e^{2k}$

$e^{2k} = 32$

$2k = \ln(32)$

$k = \frac{1}{2}\ln(32) = \ln(4\sqrt{2})$

At $t = 12$ (i.e. midnight)

$y = 5000e^{12\ln(4\sqrt{2})} = 5.369 \times 10^{12}$ (4 s.f.)

Links to Other Concepts

- Graph transformations
- Coordinate geometry
- Rates of growth and decay – modelling
- Powers and indices
- Logarithms
- Equations, solving equations
- Use of calculator

SUMMARY

- $y = a^x$ is a function that increases with x with a gradient that is always increasing.
- When $a > 1$, the graph forms a curve that passes through 1 on the y-axis and tends towards 0 as x decreases.
- e is a constant irrational number (2.71828182...).
- The gradient of the graph $y = ae^{kx}$ at any point is ake^{kx}.
- Column vectors can be used to efficiently describe translations.

QUICK TEST

1. What is the value of 4^{-2}?
2. Sketch the curve $y = 2^x$.
3. On the same axes, sketch the curve $y = \left(\frac{1}{2}\right)^x$.
4. Describe the transformation that maps the graph from question 2 onto the graph in question 3.
5. Sketch the curve $y = \left(\frac{1}{2}\right)^{(x-3)}$, labelling clearly the point $(p, 1)$ and stating the value of p.
6. What is the gradient of the curve $y = e^x$ at $x = -1$?
7. The population of a town is modelled as $P = 34\,000e^{0.02t}$, where t is the number of years after 1 January 2005.

    a) Is the population increasing or decreasing? Explain your answer.

    b) What was the population of the town on 1 January 2005?

    c) What was the population of the town on 1 January 2015?

PRACTICE QUESTIONS

1. $f(x) = 4^x$

    a) The equation $y = f(x) + a$ translates the graph so that it passes through the origin. State the value of a. **[1 mark]**

    b) The equation $y = 64 \times 4^x$ gives a curve that could be described as two transformations of $f(x)$. Describe both transformations. **[3 marks]**

    c) Sketch the graph of $y = f\left(-\frac{1}{2}x\right)$ on the same axes as the graph of $y = f(x)$. **[3 marks]**

2. The population of tigers in the wild was shown to be increasing from a census taken in 2016. An early model for the growth in the population of wild tigers was that population $= 3200e^{(0.0327)t}$, where t is the number of years after 2006.

    a) Based on this model, what was the population in 2006? **[1 mark]**

    b) What was the rate of increase in tiger population in 2016? **[2 marks]**

    c) Based on this model, what is the estimated number of wild tigers by 2050? **[2 marks]**

Logarithms

$a = b^c$

You can find a single unknown in the relationship $a = b^c$ given the other two values:

- **Finding a:** Raise b to the power c.

> **Example**
> $a = 4^{\frac{3}{2}}$ $\quad a = 2^3 = 8$

- **Finding b:** Find the c root of a.

> **Example**
> $15625 = b^6$. Find the sixth root of 15625.
> $b = \sqrt[6]{15625} = 5$

- **Finding c:** This is the **inverse** process of raising b to the power c. In words the problem is 'what power do you raise b to in order to get a?' Mathematically, this function is called the logarithm. For simple questions you may be able to find the answer from inspection and knowledge about number theory. More complex questions require a calculator.

> **Example**
> $2187 = 3^c$. Find c. $\quad c = \log_3 2187 = 7$

The graph of an inverse function is a reflection of the original function in the line $y = x$.

This graph shows the function $y = a^x$, where $a > 1$, and its inverse $y = \log_a x$ (said 'log to the base a of x').

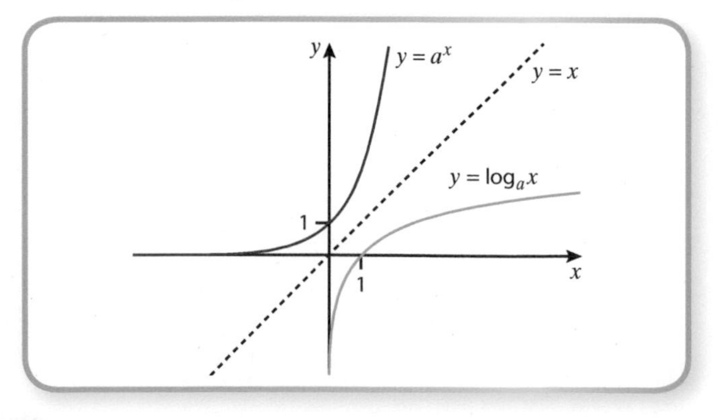

This graph shows the function $y = a^x$, where $0 < a < 1$, and its inverse $y = \log_a x$. It is generally met as a transformation of the first pair of graphs.

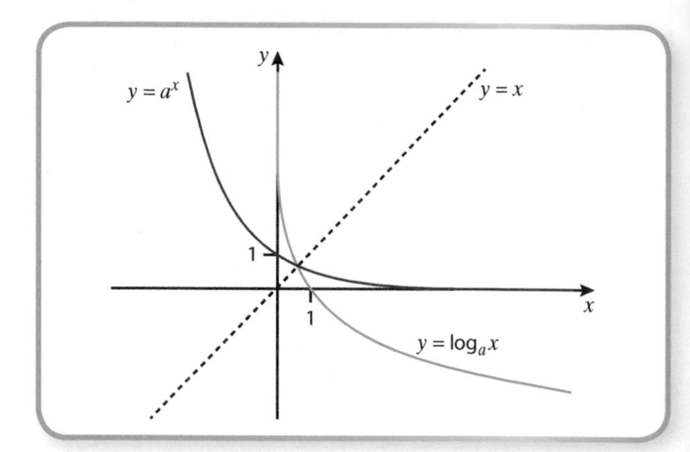

Whilst logs and exponentials can be applied as operations with negative bases, the graphs are discontinuous with many results undefined – giving a calculator error. At AS and A-level, they are defined for $a > 0$ and $x \geqslant 0$.

If $a = 1$, the graph produced would be $y = 1^x$. As $1^x = 1$ this would give a straight horizontal line passing through 1 on the y-axis.

Special Logarithms

$y = \ln x$ is the shorthand way of writing $y = \log_e x$. This is known as the natural logarithm.

$y = \log x$, where no base is specified, is $y = \log_{10} x$.

This pair of logarithms used to be the only ones available on calculators but modern calculators allow the input of different bases. The shorthand is worth noting, recognising and using; if unsure, write the full version. If your calculator doesn't input different bases, it is possible to convert the base of a logarithm (see page 49).

Transformations with the General Graph

Consider the graph $\mathbf{f}(x) = \log_p x$.

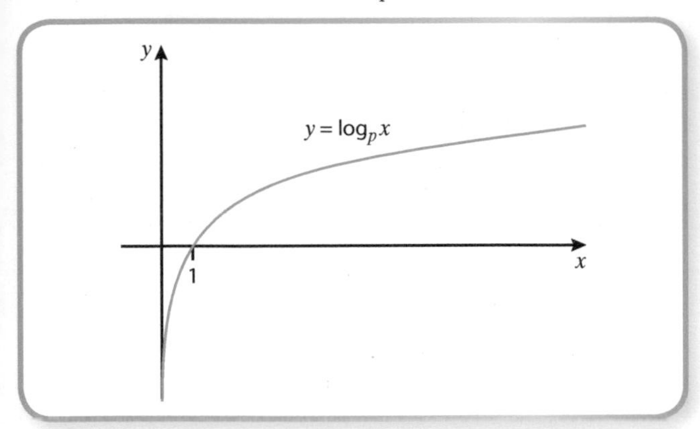

$\mathbf{f}(x) + a = \log_p x + a$

The graph is moved up by a, or translated by $\begin{pmatrix} 0 \\ a \end{pmatrix}$. If a is negative, the movement will be downwards.

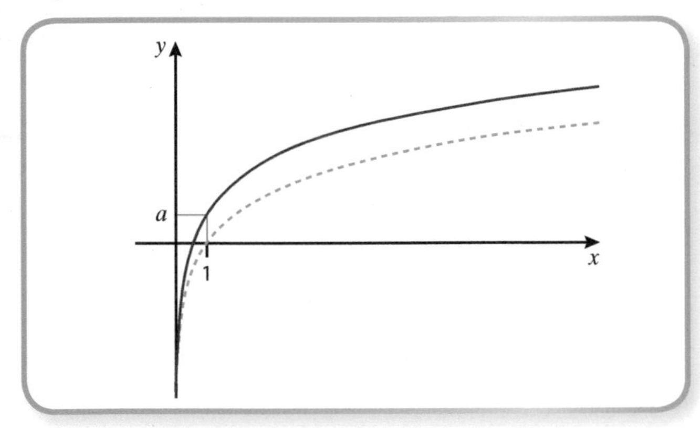

$\mathbf{f}(x + a) = \log_p (x + a)$

The graph is moved left by a, or translated by $\begin{pmatrix} -a \\ 0 \end{pmatrix}$. If a is negative, the movement will be to the right.

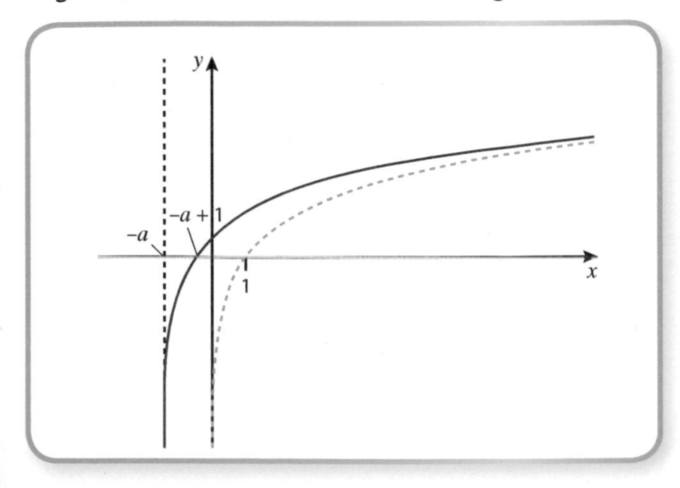

$a\mathbf{f}(x) = a\log_p x$, where a is a positive constant

The graph is stretched vertically by a. All x-intercepts stay in the same place. Each y-coordinate is $\times\ a$.

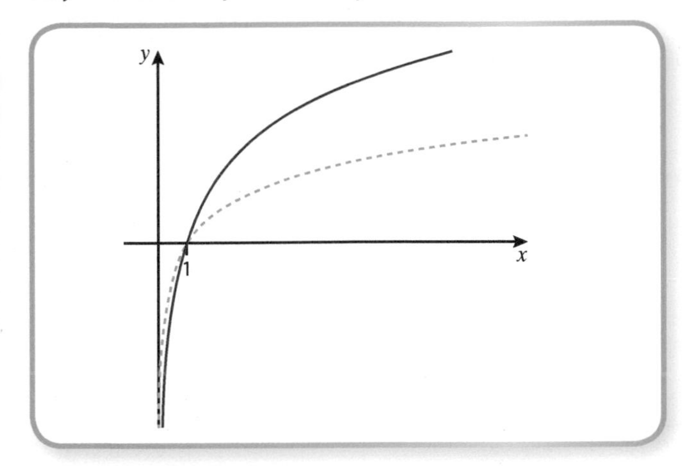

$\mathbf{f}(ax) = \log_p ax$, where a is a positive constant

The graph is stretched horizontally by $\frac{1}{a}$. The x-coordinate is $\times \frac{1}{a}$.

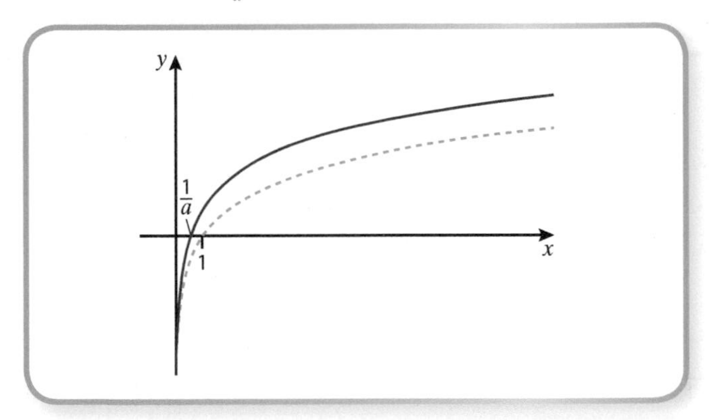

$-\mathbf{f}(x) = -\log_p x$

The graph is reflected in the x-axis.

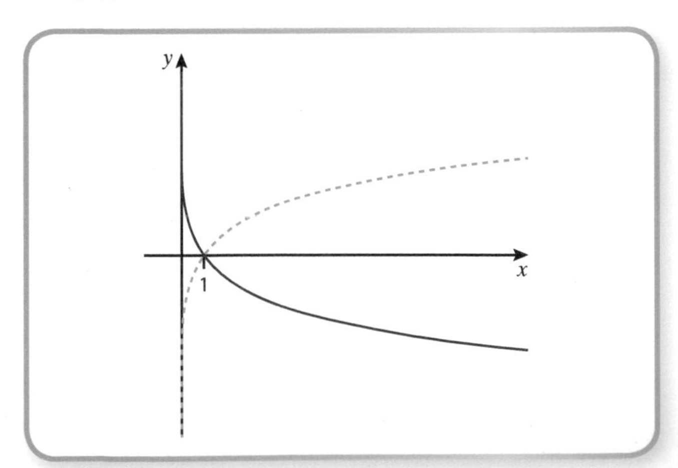

$\mathbf{f(-x) = \log_p(-x)}$

The graph is reflected in the y-axis.

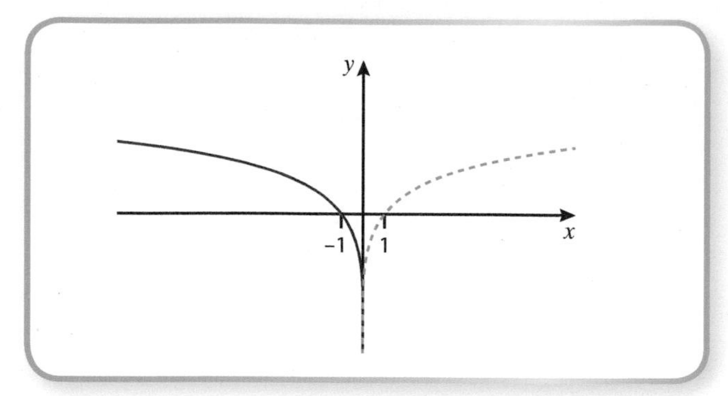

Solving and Rearranging Exponential and Logarithmic Equations

When dealing with equations using the four most common operations, it is relatively easy to follow what is happening. You can consider a simple example and apply the logic to a more complex case involving exponentials and logarithms. One way of thinking about it is modelled below.

Example

Find b given that $2+4^b=1026$.

$2+4^b=1026$

Logarithms and exponential operations fall under **I** in BIDMAS. In this case, 4 is raised to the power b then 2 is added. So to undo this, first 2 must be subtracted from both sides:

$4^b=1024$

To undo the exponential, take $\log_4$ of both sides (as $\log_4 x$ is the inverse of 4^x).

$\log_4 4^b=\log_4 1024$

The left-hand side of this equation can be read as 'to what power is 4 raised to get an answer of 4 to the power b?' The answer to which is 'b'.

$b=\log_4 1024=5$

This process is often considered as a rearrangement of the two forms of the equation, as shown below.

If... $a^b=c$ (base) 'a to the power b equals c'

...then $\log_a c=b$ 'log to the base a of c equals b'

Example

Find x when $\frac{2^{x+3}+1}{3}=2731$.

Step 1: Multiply both sides by 3.

$\frac{2^{x+3}+1}{3}\times 3=2731\times 3$

$2^{x+3}+1=8193$

Step 2: Subtract 1 from both sides.

$2^{x+3}+1-1=8193-1$

$2^{x+3}=8192$

Step 3: Take log to the base 2 of both sides. (The $x+3$ falls within the 'brackets' in BIDMAS.)

$\log_2 2^{x+3}=\log_2 8192$

$x+3=13$

Step 4: Subtract 3.

$x=10$

In questions that ask for exact values, answers should be left in terms of the logarithm if necessary.

Example

Find the exact value of p when $e^{p+1}=7$.

$\ln e^{p+1}=\ln 7$

$p+1=\ln 7$

$p=\ln 7-1$

If a question is a logarithm, with the unknown enclosed, then the inverse operation to use is the exponential with the base given by the logarithm.

Example

Find the exact value of t when $3\ln t+2=47$.

Subtract 2 from each side: $3\ln t=45$

Divide each side by 3: $\ln t=\frac{45}{3}=15$

Having isolated the logarithm, exponentials are used on both sides. In this case the base is e, as it is the 'natural logarithm' ln.

$e^{\ln t}=e^{15}$

As $e^{\ln t}=t$, $t=e^{15}$

Links to Other Concepts

- Exponentials
- Graph transformations
- Coordinate geometry
- Rates of growth and decay – modelling
- Powers and indices
- Solving equations

SUMMARY

- **Logarithms are the inverse functions of exponentials.**
- **The graph of $\log_a x$ is the reflection of $y = a^x$ in the line $y = x$. The graph passes through 1 on the x-axis and when $a > 1$ tends towards the negative y-axis as x approaches 0.**
- **$\ln x = \log_e x$**
- **$\log x = \log_{10} x$**
- **$a^{\log_a b} = b$ and $e^{\ln b} = b$**
- **$\log_a(a^b) = b$ and $\ln(e^b) = b$**
- **$\log_a a = 1$ and $\ln(e) = 1$**
- **$\log_a 1 = 0$, assuming $a \neq 1$, and $\ln(1) = 0$**
- **If a question asks for an exact answer, it can be left in terms of the logarithm or the exponential.**

QUICK TEST

1. $e^{x+1} = e^{2x-3}$. Find x.
2. Evaluate $\ln e^7$.
3. Evaluate $\log_3 81$.
4. Make q the subject of $a^{q-2} + 5 = b$.
5. Sketch the graph of $y = \ln(3x)$, showing the axis-intercepts.
6. $b = \log_2\left(\frac{1}{8}\right)$. Find b.
7. Sketch the graph of $y = 2 + \ln(x)$.
8. $10\log_4 b = 5$. Find b.
9. $4e^{x-1} + 1 = 29$. Find the exact value of x.
10. Sketch the graph of $y = -2\ln x$.

PRACTICE QUESTIONS

1. It is given that $\log_4 b = a$.

a) Write b in terms of a. **[1 mark]**

b) Hence or otherwise, find the value of $\log_2 b$ in terms of a. **[3 marks]**

2. The graph shows the curve $y = \ln(px)$.

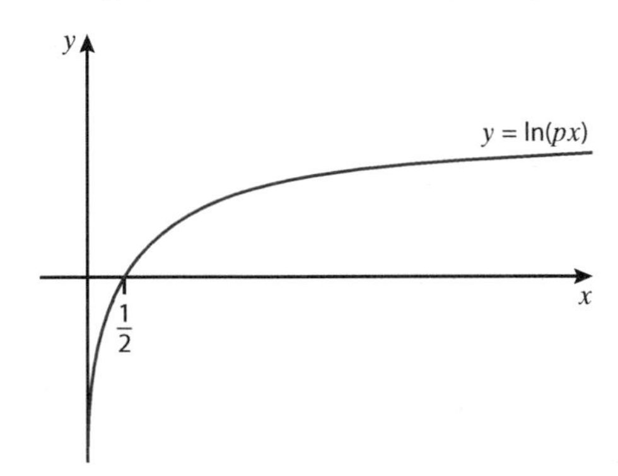

a) State the value of p. **[1 mark]**

b) On the same axes, sketch the curve $y = 3\ln(2x)$. **[2 marks]**

c) Find the exact x-value for the curve $y = 3\ln(2x)$ at the point when $y = -15$. **[4 marks]**

3. The line $y = 25$ intersects the curve $y = 3e^{2x+1} - 5$ at point P. The point where the graphs meet has coordinates $((a(\ln 10) + b), c)$.

Find the values of a, b and c. **[6 marks]**

Log Rules

In the previous topic, the logs were introduced as a function and operation. Sometimes **equations** involving logarithms are more complex (e.g. there may be two, or more, terms written as logs or **indices** within the equation). As long as the base is the same, you can combine them in some circumstances. You can also change the **base** of a logarithm so that combinations are possible. This is closely related to the index laws but feels much less intuitive.

Reminder of the Index Laws

If the base is the same, then:

$x^a \times x^b = x^{a+b}$ $x^a \div x^b = x^{a-b}$

$(x^a)^b = x^{ab}$ $x^{-a} = \frac{1}{x^a}$

$x^{\frac{a}{b}} = \sqrt[b]{x^a} = (\sqrt[b]{x})^a$

If two indices have different bases, you can change the base so that they are the same. Look for numbers that are related as repeat additions.

Example

Simplify $4^3 \times 2^7 \times 16$.

As each number can be written as a power of 2, the above expression could be written as:

$(2^2)^3 \times 2^7 \times (2^4) = 2^6 \times 2^7 \times 2^4 = 2^{6+7+4} = 2^{17}$

You could use any power of 2 in the base (2, 4, 8, 16, ...). By using 2 (the smallest) it gives integer powers. Using a base of 4 would give some fractional powers.

Example

$4^3 \times \left(4^{\frac{1}{2}}\right)^7 \times 4^2 = 4^{\frac{17}{2}}$

If given the choice, it is advisable to use the simplest numbers. However a question will usually state the form it requires.

Introducing the Log Rules

Logarithms also rely on the same base in order to be able to combine the terms. Linking them back to the index laws can help to understand them. This is done for each rule below.

$\log_c a + \log_c b = \log_c ab$

Let $x = \log_c a$ and $y = \log_c b$.

Write in index form: $a = c^x$ and $b = c^y$

$c^x \times c^y = a \times b$

$c^{x+y} = ab$

Take log base c of both sides: $\log_c c^{x+y} = \log_c ab$

$x + y = \log_c ab$

Substitute back in for x and y: $\log_c a + \log_c b = \log_c ab$

Example

$\log_4 16 + \log_4 4 = \log_4 64$

By considering these relatively simple numbers, you can see the rule at work. Having a simple example to check is a good way to help remember a rule, if needed.

$\log_4 16 + \log_4 4 = 2 + 1 = 3$

$\log_4 64 = 3$

$\log_c a - \log_c b = \log_c \frac{a}{b}$

Let $x = \log_c a$ and $y = \log_c b$.

Write in index form: $a = c^x$ and $b = c^y$

$c^x \div c^y = a \div b$

$c^{x-y} = \frac{a}{b}$

Take log base c of both sides: $\log_c c^{x-y} = \log_c \frac{a}{b}$

$x - y = \log_c \frac{a}{b}$

Substitute back in for x and y: $\log_c a - \log_c b = \log_c \frac{a}{b}$

Example

$\log_2 16 - \log_2 4 = \log_2 \frac{16}{4} = \log_2 4 = 2$

$\log_2 16 - \log_2 4 = 4 - 2 = 2$

$\log_c a^n = n\log_c a$

Let $x = \log_c a \leftrightarrow a = c^x$

Write in index form: $a^n = \left(c^x\right)^n$

$a^n = c^{nx}$

Take log base c of both sides: $\log_c a^n = \log_c c^{nx}$

$\log_c a^n = nx$

Substitute back in for x: $\log_c a^n = n\log_c a$

Example

$\log_5 5^3 = 3\log_5 5$

$\log_5 5^3 = 3$

$3\log_5 5 = 3 \times 1 = 3$

When using the first two rules (addition and subtraction), there is no coefficient in front of the log. The third rule enables you to convert between a coefficient in front of the logarithm and a power contained within the logarithm, which might be necessary before the addition and subtraction rules can be used.

Example

Express $5\log_7 2 + \frac{1}{2}\log_7 8$ as a single logarithm.

$5\log_7 2 + \frac{1}{2}\log_7 8 = \log_7 2^5 + \log_7 8^{\frac{1}{2}}$

$= \log_7 32 + \log_7 (2\sqrt{2})$

$= \log_7 (32 \times 2\sqrt{2})$

$= \log_7 64\sqrt{2}$

As shown in this example, there may be a root or index within the logarithm. There could also be algebraic terms within the logarithm or as the base. Don't be put off by this!

Converting the Base of a Logarithm

In the days of log tables and less capable technology, being able to convert between bases for logs was vital as only certain logs were represented. Modern calculators can deal with any base. However, it is still useful to convert the base of a log when exam questions include algebra or ask for working to be 'shown'.

$\log_a b = \frac{\log_c b}{\log_c a}$

Let $x = \log_a b \leftrightarrow b = a^x$

$\log_c b = \log_c a^x$

$\log_c b = x\log_c a$

$x = \frac{\log_c b}{\log_c a}$

$\log_a b = \frac{\log_c b}{\log_c a}$

Example

Convert the base of the logarithm $\log_3 5$ to e.

$\log_3 5 = \frac{\log_e 5}{\log_e 3} = \frac{\ln 5}{\ln 3}$

As with converting bases within index form, it is relatively straightforward if the bases of the logarithms to be combined are powers of each other.

Example

Write $\log_2 7 + \log_4 3$ as a single logarithm in the form $\log_4 q$.

First convert the bases to 4, as directed by the question:

$\log_2 7 = \frac{\log_4 7}{\log_4 2}$

Given that $\log_4 2 = \frac{1}{2}$ (as $4^{\frac{1}{2}} = 2$):

$\log_2 7 = \frac{\log_4 7}{\frac{1}{2}} = 2\log_4 7 = \log_4 7^2$ (as $a\log_b c = \log_b c^a$)

$\log_2 7 + \log_4 3 = \log_4 7^2 + \log_4 3$

$= \log_4 49 \times 3 = \log_4 147$

Solving More Complex Equations Using the Log Laws

Pulling the laws together means more complicated equations can be solved. The aim is always to get all the logarithms together, if possible. If there is an equivalent log on each side, they can be raised as the power of the base to find the answer. For this to happen, all the values on one side must be within the log. Any coefficients need to be simplified as well.

Example

$\log_3 8a = \log_3 2 + \log_3 24 - \log_3 \frac{a}{6}$. Find the value of a, a positive integer.

Ensuring you comply with BIDMAS, there are different ways of approaching this question. Two are shown here.

Method 1:

$\log_3 8a = \log_3 2 + \log_3 24 - \log_3 \frac{a}{6}$

$\log_3 8a = \log_3 \left(2 \times 24 \div \frac{a}{6}\right)$

$\log_3 8a = \log_3 \frac{288}{a}$

$8a = \frac{288}{a}$

$a^2 = 36$

$a = \sqrt{36} = 6$

Method 2:

$\log_3 8a = \log_3 2 + \log_3 24 - \log_3 \frac{a}{6}$

$\log_3 8a + \log_3 \frac{a}{6} = \log_3 2 + \log_3 24$

$\log_3 \left(8a \times \frac{a}{6}\right) = \log_3 (2 \times 24)$

$\log_3 \left(\frac{4a^2}{3}\right) = \log_3 48$

$\frac{4a^2}{3} = 48$

$4a^2 = 48 \times 3 = 144$

$a^2 = 144 \div 4 = 36$

$a = \sqrt{36} = 6$

Note: Given that 3^n is positive, $8a$ must be positive and as such we can ignore the -6.

With greater confidence, you should be able to manipulate logarithms using the laws both ways.

Example

The expression $2\log_3 18$ can be written in the form $a + \log_3 a$. Find the value of a.

There are different ways of approaching this question. One possible solution is shown here. Whatever first step is taken, the result will be the same if the laws are followed correctly.

$2\log_3 18 = 2\log_3 (9 \times 2)$

$= 2(\log_3 9 + \log_3 2)$

$= 2(2 + \log_3 2)$

$= 4 + 2\log_3 2$

$= 4 + \log_3 2^2$

$= 4 + \log_3 4$

$a = 4$

Example

$y = \ln 12x + \log 10 - \ln 4$

a) Express y in terms of $\ln(\mathrm{f}(x))$.

$y = \ln(12x \div 4) + 1$

(since $\log_a a = 1$, $\log 10 = 1 = \ln e$)

$y = \ln(3x) + \ln e$

$y = \ln(3ex)$

b) Hence express x in terms of y.

Take exponentials of each side:

$e^y = e^{\ln(3ex)}$

Using the law $e^{\ln a} = a$:

$e^y = 3ex$

Rearrange to isolate x:

$x = \frac{e^y}{3e}$

Links to Other Concepts

- Exponentials
- Coordinate geometry
- Rates of growth and decay – modelling
- Powers and indices
- Equations, solving equations
- Quadratics

SUMMARY

- **Logarithms are the inverse functions of exponentials.**
- $a^{\log_a b} = b$
- $\log_a(a^b) = b$

$\log_a a = 1$

$\log_a 1 = 0$, **assuming** $a \neq 1$

If a question asks for an exact answer, it may be left in terms of the logarithm or exponential (where it doesn't give a simple numerical answer).

The laws for combining logs:

$\log_c a + \log_c b = \log_c ab$

$\log_c a - \log_c b = \log_c \frac{a}{b}$

$\log_c a^n = n\log_c a$

The laws when working in base e:

$\ln a + \ln b = \ln ab$

$\ln a - \ln b = \ln\left(\frac{a}{b}\right)$

$\ln a^n = n\ln a$

The laws when working in base 10:

$\log a + \log b = \log ab$

$\log a - \log b = \log\frac{a}{b}$

$\log a^n = n\log a$

Converting the base: $\log_a b = \frac{\log_c b}{\log_c a}$

The laws for combining logarithms feel fairly counter-intuitive, especially when first met. Take your time answering questions that involve them and check your answers if possible (substitute your answer back into the original question, for example).

QUICK TEST

1. Write the expression $\log_2 128 = 7$ in index form.
2. Write the expression $4^x = 63$ in log form.
3. Find the missing values in each equation:
   a) $\log_a 4 + \log_a 3 = \log_a$ ◆
   b) $\log_a$ ◆ $+ \log_a 7 = \log_a 14$
   c) $\log_a 15 - \log_a 3 = \log_a$ ◆
   d) $\log_2 18 - \log_2$ ◆ $= \log_2 6$
   e) $5\log_a 4 = \log_a$ ◆
   f) ◆$\log_5 2 = \log_5 64$
4. Write the following in the form given:
   a) $\log_9 12$ to base 3
   b) $\log_2 15$ to base 4
   c) $\ln 4$ to base 2
5. $\log_2 24$ can be written in the form $a + \log_2 a$. Find a.
6. Write $\log_a(b\sqrt{c})$ in terms of $\log_a b$ and $\log_a c$.

PRACTICE QUESTIONS

1. Write $\log_9 12$ in the form $\frac{1}{2}(1 + \log_3 b)$. **[3 marks]**
2. a) Show that $(3^x - 12)(3^x - 3) = 3^{2x} - 5 \times 3^{x+1} + 36$. **[3 marks]**
   b) Hence find the exact solutions to the equation $3^{4x} - 5 \times 3^{2x+1} + 36 = 0$. **[4 marks]**
3. $4 - \log_2(x - 2) = \log_2(x + 13)$. Find the value of x. **[6 marks]**

Using Logarithms

Estimating Parameters

Polynomial relationships (of the form $y = ax^n$) and exponential relationships (of the form $y = kb^x$) are two common non-linear relationships. By using logarithms it is possible to view this relationship as linear and hence find unknown values. You can plot the relationship between x and y using a logarithmic scale. You can also rewrite the relationship in terms of a logarithm to give it in the form of a straight line: $y = mx + c$.

$y = ax^n$

Using the log rules from the last topic and by taking logs of both sides gives:

$\log y = \log ax^n$

$\log y = \log x^n + \log a$

$\log y = n\log x + \log a$

Let $Y = \log y$, $X = \log x$ and $c = \log a$: $\boldsymbol{Y = nX + c}$

This is now an equation in the form $y = mx + c$, which is that of a straight line. As such, it is possible to use the linear relationship and given values to find the unknown values.

Example

The relationship between two variables is defined as $y = ax^n$, where a and n are constants. Use the data in the table to find the relationship between x and y, having found the constants a and n.

x	2	3	4	5	6
y	52	175.5	416	812.5	1404

Take logs (to base 10) of both sides
$\rightarrow \log y = \log ax^n$

$\log y = n\log x + \log a$

$Y = nX + c$ is now a linear relationship between Y and X.

Use $Y = \log y$ and $X = \log x$ to find values for the linear relationship.

$X = \log x$	log 2	log 3	log 4	log 5	log 6
$Y = \log y$	log 52	log 175.5	log 416	log 812.5	log 1404

The graph is a straight line with gradient n and y-intercept of c.

To find the gradient, take two pairs of values from the X, Y table (gradient is change in Y divided by change in X).

Gradient, $n = \frac{\log 416 - \log 52}{\log 4 - \log 2} = 3$

$\boldsymbol{n = 3}$ (found using a calculator but could be simplified to $\frac{\log\left(\frac{416}{52}\right)}{\log\left(\frac{4}{2}\right)} = \frac{\log 8}{\log 2} = 3$)

$Y = 3X + c$

$\log 52 = 3\log 2 + c$

$c = \log 52 - 3\log 2$

Use the log laws to combine the logs. This could be done on a calculator, intermediate step $c = 0.81291\ldots$

$c = \log 52 - \log 2^3$

If using a calculator, use the memory to maintain the full accuracy of the answer, then use 10^c to find a.

$= \log \frac{52}{8} = \log 6.5$

$c = \log a = \log 6.5 \quad \rightarrow \quad a = \mathbf{6.5}$

$\boldsymbol{y = 6.5x^3}$

$y = kb^x$

Using the log rules and by taking logs of both sides:

$\log y = \log kb^x$

$\log y = \log b^x + \log k$

$\log y = x\log b + \log k$

Let $Y = \log y$, $m = \log b$ and $c = \log k$: $\boldsymbol{Y = mx + c}$

This is now an equation in the form $y = mx + c$. As such, it is possible to use the linear relationship and given values to find the unknown values.

Example

The relationship between two variables is defined as $y = kb^x$, where b and k are constants. Use the data in the table to find the relationship between x and y, having found the constants b and k.

x	5	10	15	20
y	48	1536	49 152	1 572 864

Taking logs of both sides:

$\log y = \log kb^x$

Use laws of logs to separate the unknowns on the right-hand side:

$\log y = \log b^x + \log k = x\log b + \log k$

$Y = mx + c$, where $Y = \log y$, $m = \log b$ and $c = \log k$, giving updated linear relationship between x and Y:

x	5	10	15	20
$Y = \log y$	log 48	log 1536	log 49 152	log 1 572 864

Gradient, $m = \frac{\log 1536 - \log 48}{10 - 5}$

$= \frac{\log \frac{1536}{48}}{5} = \frac{1}{5}\log 32$

$= \log 32^{\frac{1}{5}} = \log 2$

As $m = \log b$ $\qquad$ $\boldsymbol{b = 2}$

$Y = mx + c$

$\log 48 = \log 2 \times 5 + c$

$c = \log 48 - \log 32$

$c = \log \frac{48}{32} = \log \frac{3}{2}$ $\qquad$ $\boldsymbol{k = \frac{3}{2} = 1.5}$

Therefore $\boldsymbol{y = 1.5 \times 2^x}$

Real-Life Data

The previous examples are both based on a strict exponential or polynomial relationship between x and y. In real life, when conducting scientific measurements, there is always a margin of error in every reading. For this reason the points will not plot perfectly into a straight line (having used logs to gain the linear relationship) as the previous examples would. It is possible that, given a set of results, a line of best fit is the best method to find the linear relationship and generate the unknown values to define the relationship between x and y.

Example

The velocity of water passing through a pipe is modelled as constant. By using different radii of pipes, Emily takes measurement of the volumetric flow rate produced (i.e. the volume of water through the pipe every unit of time). She models the relationship as $y = ax^n$. Her table of results is shown.

a) Complete the table of results.

Radius	x	2	3	4	5	6
Volumetric flow rate	y	97.7	166.0	407.4	562.3	758.6
Plotted to accuracy of 3 s.f.	$X = \log x$	0.301	**0.477**	0.602	0.699	0.778
	$Y = \log y$	1.99	**2.22**	2.61	2.75	2.88

b) Plot the missing point on the graph and use it to estimate the values of a and n. Show your working clearly.

Plotting the missing point and drawing a line of best fit:

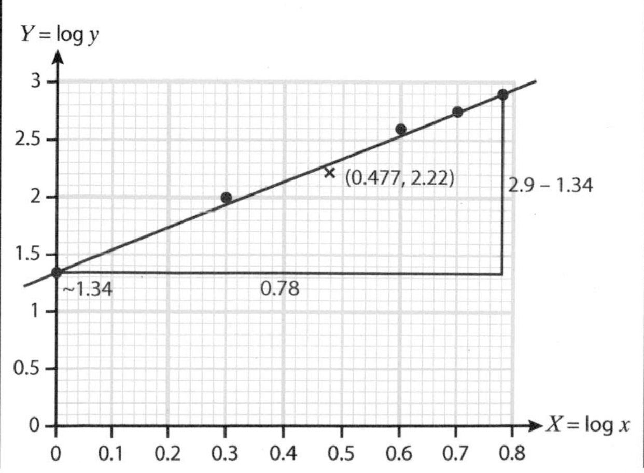

y-intercept ≈ 1.34
(the reading must be accurate to the graph)

$c = \log a$

$1.34 = \log a$

$a = 10^{1.34} = 21.9$ (3 s.f.)

Gradient of the line, $n = \frac{2.9 - 1.34}{0.78} = 2$

Note: Working on the graph counts as 'showing clear working', as long as it is obvious where values have come from and what has been done.

Exponential Growth and Decay

Growth and decay modelled using exponentials (and therefore logarithms) occurs in many real-life contexts.

Equations are of the form $y = Ae^{bx} + C$ or $y = Ak^{t} + C$.

The difference from the last set of examples is that the equations are slightly more complex, and may not reduce easily to a linear relationship. However, if enough data is available, unknowns can still be found using simultaneous equations. Quite often the information for $t = 0$ is given. Using the fact that anything to the power 0 is 1, this would simplify both equations above to $A + C$.

Questions may ask for unknown values, the size of population after a certain length of time (for example), or for an analysis of the model used.

If asked about the model, think about the long-term effect. Is the value of y able to increase (or decrease) indefinitely? Is there enough information to base the modelling assumptions upon?

Example

The population of a species of beetle inhabiting an island is being monitored. The population N after t years, from the start of the study, is modelled as $N = Ae^{kt} + 1500$. At the start of the study it was estimated that there were 2000 beetles. After eight years the population is 1601.

a) Find the value of A.

At $t = 0$: $2000 = A \times e^{k \times 0} + 1500$

$2000 - 1500 = A \times 1$

$A = 500$

b) What size is the population tending towards?

As the population is declining, k is negative.

As $t \to \infty$, $Ae^{kt} \to 0 \ \therefore N \to 1500$ beetles

c) What was the size of the population after four years?

First find k using the information for eight years:

$1601 = 500e^{k \times 8} + 1500$

$1601 - 1500 = 500e^{8k}$

$e^{8k} = \frac{101}{500}$

$8k = \ln\frac{101}{500} = -1.5994875\ldots$

$k = -0.1999359\ldots = -0.200$ (3 s.f.)

Use the values for A and k to find N at four years:

$N = 500e^{-0.2 \times 4} + 1500 = 1724.6644\ldots$

After four years the population was approximately 1725 beetles.

SUMMARY

- **Log graphs can be used to estimate parameters in relationships in the form $y = ax^n$ and $y = kb^x$ if data is given for x and y.**
- **By converting the equation $y = ax^n$, it is possible to rewrite in the form $Y = nX + c$, where $Y = \log y$, $X = \log x$ and $c = \log a$.**
- **By converting the equation $y = kb^x$, it is possible to rewrite in the form $Y = mx + c$, where $Y = \log y$, $m = \log b$ and $c = \log k$.**
- **Scientific experiments are likely to give imperfect data so lines of best fit can be used to find the trend and model it.**
- **For exponential growth and decay problems, use given conditions to find the missing information.**
- **If asked to consider if the model is realistic, consider long-term implications.**

Links to Other Concepts

- Exponentials
- Coordinate geometry
- Rates of growth and decay – modelling
- Powers and indices
- Solving equations
- Population size, spread of disease, half-lives

QUICK TEST

1. What is the value of ln32 to 3 significant figures?
2. The equation $y = ax^n$ is rewritten in the form $Y = nX + c$. Write the equations for Y, X and c in terms of y, x and a.
3. The equation $y = 6 \times 3^x$ is rewritten in the form $Y = mx + c$ and plotted on a graph. What is the y-intercept?
4. The graph shows the line $Y = mx + c$, where $Y = \log y$, $m = \log b$ and $c = \log k$.

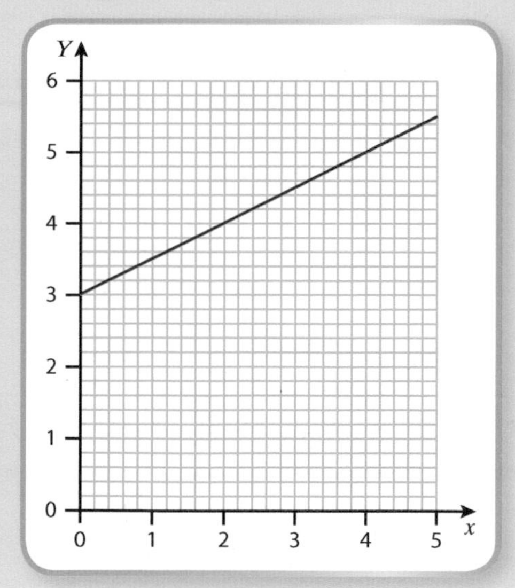

   a) Find the value of k.

   b) Find the value of b.

   c) Find the value of y when $x = 4$.

5. The radioactive half-life of carbon-14 is 5730 years. The initial mass of carbon in a sample is 200 g. The mass (m) after t years is given by the equation $m = A\left(\frac{1}{2}\right)^{\frac{t}{5730}}$

   a) What is the value of A?

   b) What is the mass after 2000 years?

PRACTICE QUESTIONS

1. Miriam is sterilising drinking water by boiling. She boils a pan of water, then leaves it outside to cool down. At any given time (t) in minutes, the temperature (T) of the water is given by the equation $T = Ae^{-kt} + C$. The temperature outside (C) is a constant 12 °C. The water boils at 100 °C.

   a) Find the value of A. **[2 marks]**

   b) After 4 minutes the water has cooled to 89 °C. Find the exact value of k. **[3 marks]**

   c) Hence find the length of time it takes for the water to fall below 25 °C. **[6 marks]**

2. A relationship between two variables is of the form $y = ax^n$. The graph below shows the linear relationship created by taking $\log_2$ of each side.

   a) Show that the equation can be written in the form $Y = nX + c$, stating the substitutions made for Y, X and c.

   [3 marks]

   The graph shows the plot of the line $Y = nX + c$.

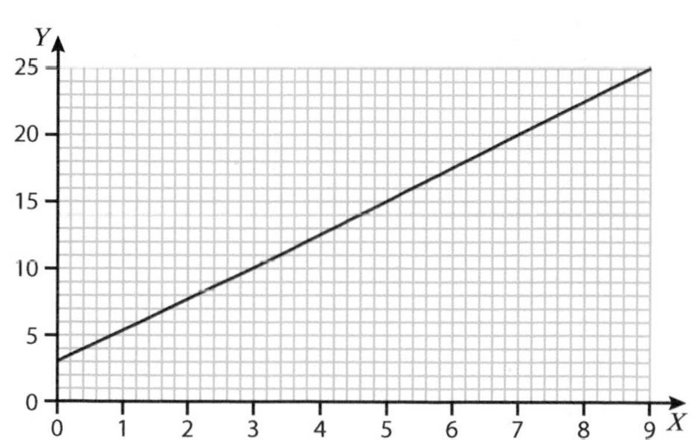

   b) Find the values of a and n. **[4 marks]**

   c) Hence find the value of y when $x = 100$.

   [2 marks]

Differentiation 1

Differentiation and its inverse (integration) form the key components of differential calculus. Calculus is the study of continuous change.

Differentiation is a method of finding the gradient function.

To find the gradient of a straight line, use $\frac{\text{change in } y}{\text{change in } x}$. As the gradient is constant, this can be done with any pair of coordinates and the result will be the same.

Tangents and Gradient

On a curve, the gradient can be approximated using a tangent (a line which touches the curve but does not cross it). By drawing an 'envelope' of these tangent lines, the curve could be considered to be made up of an infinite number of infinitesimally small straight line segments. Each tangent represents the gradient at just one point. If asked to approximate or estimate a gradient from a graph, draw in the tangent by eye then calculate the gradient of the tangent.

Example

Approximate the gradient of the curve at $x = 4$.

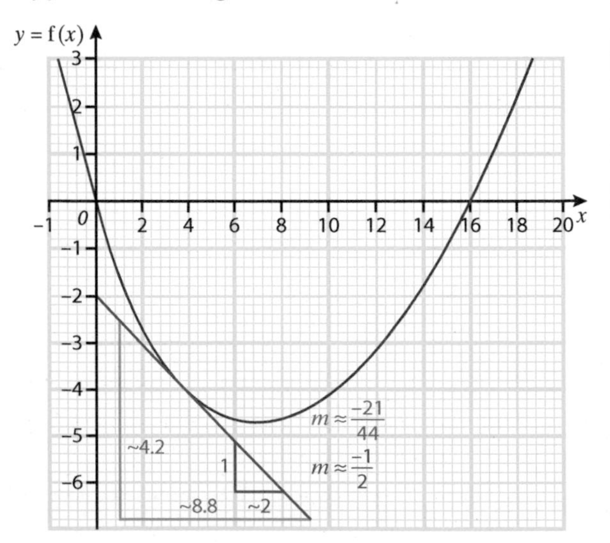

Answers will vary as drawing the tangent line isn't precise. However, values should fall within a range of tolerance. Show the steps taken on the diagram or with clear working below.

When finding the gradient of the line, look for where the line passes through clearly defined points.

The larger the triangle used to calculate the gradient, the smaller the effect of residual errors. In this example, the scale means that the accuracy of the readings is limited. As a general rule, take the largest triangle with as clearly defined points as possible.

Here, two possible answers are found for the gradient. The larger triangle is likely to have given more accurate results, but both are close to each other and, as the question asks for an approximation, either is acceptable.

Watch out for positive (/) and negative (\) gradients.

Differentiation from First Principles

The diagram shows a polynomial curve $y = f(x)$. The tangent to the curve at point (x, y) is drawn on. By taking the straight line between two points on the curve, you can approximate the gradient of the tangent.

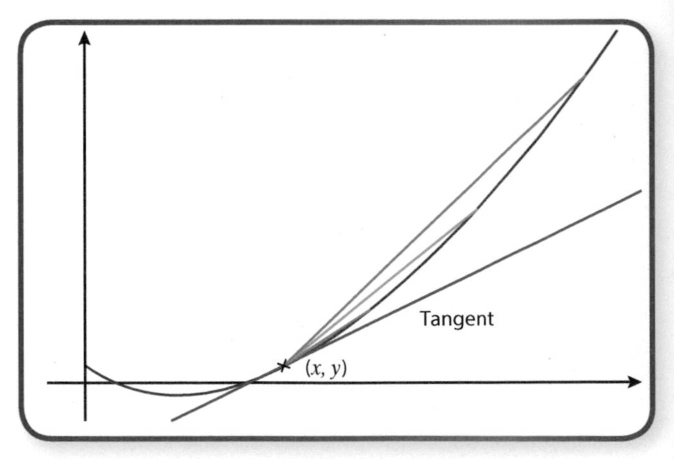

As those two points get closer and closer together, the gradient of the line produced gets closer to the gradient of the tangent.

This idea is represented by $\frac{\delta y}{\delta x}$, which means a very small change in y divided by a very small change in x.

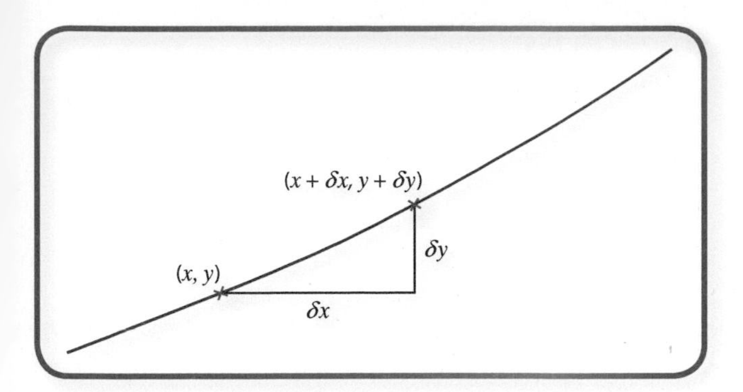

It is important to be able to differentiate from first principles for small positive powers of x. In the example below a cubic curve is used, but the principle and process would apply to any simple polynomial.

Example

Differentiate $y = x^3$ from first principles.

A point (x, y) is taken on the curve. A second point is taken very close to the first, with coordinates $(x+\delta x, y+\delta y)$.

$y+\delta y=(x+\delta x)^3$

Expand the bracket using binomial expansion or by multiplying out brackets:

$$=x^3+3x^2(\delta x)+3x(\delta x)^2+(\delta x)^3$$

$$\delta y=3x^2(\delta x)+3x(\delta x)^2+(\delta x)^3$$

The gradient is $\frac{\delta y}{\delta x}=\frac{3x^2(\delta x)+3x(\delta x)^2+(\delta x)^3}{\delta x}$

$$=3x^2+3x(\delta x)+(\delta x)^2$$

Gradient of the tangent $\frac{dy}{dx}=\lim_{\delta x\to 0}\frac{\delta y}{\delta x}$

$=\lim_{\delta x\to 0} 3x^2+3x(\delta x)+(\delta x)^2=3x^2$

The above statement translates into English as: the gradient of the line is equal to the limit of $\frac{\delta y}{\delta x}$ as delta x tends towards 0, which is the limit of the function $3x^2+3x(\delta x)+(\delta x)^2$ as delta x tends towards 0. As the equation approaches this limit, terms that contain the delta x, e.g. $3x(\delta x)$, tend towards 0 $\left(\lim_{\delta x\to 0} 3x(\delta x)\to 3x\times 0=0\right)$. You need to be able to express and interpret limits mathematically.

Algebraic Differentiation of Polynomials

Differentiation is an algebraic method used to find the gradient function $f'(x)$ of a given function $f(x)$.

If considering a single term from a polynomial, i.e. ax^b, the differential of the term is $(ab)x^{b-1}$. In other words 'multiply the coefficient by the index then subtract 1 from the index'.

Example

Find the gradient function of $y=2x^5$.

$\frac{dy}{dx}=5\times 2x^{5-1}=10x^4$

Special cases to watch out for are when $b = 1$ and $b = 0$, as they are often written without a power.

$ax=ax^1$ $\qquad$ $\frac{d}{dx}ax^1=ax^0=a\times 1=a$

Example

$y = 8x$

$\frac{d}{dx}8x=1\times 8x^0=8\times 1=8$

$y = 8x$ produces a straight-line graph and the gradient of a straight line is constant so, whatever the x-value, the gradient of this line will always be 8.

$a = ax^0$ $\qquad$ $\frac{d}{dx}ax^0=0x^{-1}=0$

Example

$y = 12$

$\frac{d}{dx}12=0\times 12x^{-1}=0$

Again consider the graph to see why the result makes sense. $y = 12$ is a straight horizontal line. A horizontal line has gradient 0.

When a polynomial has multiple terms, each term is differentiated independently and the answer is the sum of the differentiated terms.

Example

Differentiate $y=3x^2-5x+2$.

$\frac{dy}{dx}=2\times 3x^{2-1}-1\times 5x^{1-1}+0\times 2=6x-5$

The reason the constant (+2) doesn't affect the gradient function can be linked to the graph transformations. The graphs shown are vertical translations of each other, but at each x-value the gradient of the graphs is the same.

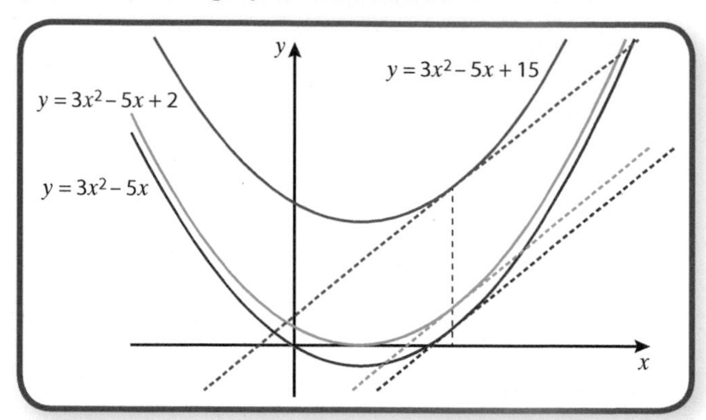

Non-Integer Indices

All the examples above have had integer powers. The method holds for any power of x: positive, negative, fractional. A question can appear harder when the index is hidden within alternate notation. The first step in many questions is to expand brackets and write each term in simplified index form.

Example

Find the gradient function of $y=(\sqrt{x}+2)\left(\frac{1}{x}-2\right)$.

First, either expand the brackets or convert to index form:

$$(\sqrt{x}+2)\left(\frac{1}{x}-2\right)=\frac{1}{\sqrt{x}}+\frac{2}{x}-2\sqrt{x}-4$$

Using the index laws, this can be rewritten to make it easy to differentiate:

$$=x^{-\frac{1}{2}}+2x^{-1}-2x^{\frac{1}{2}}-4$$

Now that each term is in index form, each of them is multiplied by the power, then has 1 taken from the power:

$$\frac{dy}{dx}=-\frac{1}{2}\times x^{\left(-\frac{1}{2}-1\right)}+(-1)\times 2x^{(-1-1)}-\frac{1}{2}\times 2x^{\frac{1}{2}-1}-0\times 4$$

This level of working doesn't need to be shown as it is implied by the next line. However, more detail can help to avoid errors, especially with negatives.

$$=-\frac{1}{2}x^{-\frac{3}{2}}-2x^{-2}-x^{-\frac{1}{2}}$$

Note on Notation

$\mathbf{f}'(x)$ $\quad \frac{dy}{dx}$ $\quad \frac{d}{dx}\mathbf{f}(x)$

These all mean the differential (or derivative) of a function, in the same way that being asked to 'draw the graph of $y=3x^2+2$' is the same as being asked to 'draw the graph of $y=f(x)$, where $f(x)=3x^2+2$'. The function notation $f(x)$ and $f'(x)$ can be simpler to write. If $f(x)$ isn't defined within the question, then it should be defined before using it.

Example

$y=4x^2+3x$. Find the gradient function.

Let $f(x)=4x^2+3x$, then $f'(x)=8x+3$.

Finding the Gradient at a Point

Having found the gradient function, the gradient can be found for any given x-value by substituting it in. It could also be used to find the coordinates of a point on the graph with any given gradient.

Example

Given that $f(x)=6x^{\frac{4}{3}}+4x-2$:

a) Find the value of $f'(x)$ when $x=-8$.

$f'(x)=8x^{\frac{1}{3}}+4$

At the point where $x=-8$:

$f'(-8)=8(-8)^{\frac{1}{3}}+4=-12$

b) Find the coordinates of the point on the curve, $f(x)$, for which $f'(x)=10$.

$f'(x)=8x^{\frac{1}{3}}+4$

$8x^{\frac{1}{3}}+4=10$

$8x^{\frac{1}{3}}=6 \quad \therefore \quad x^{\frac{1}{3}}=\frac{3}{4}$

$x=\frac{27}{64}$

To find the y-coordinate, substitute back into $f(x)$:

$6\times\left(\frac{27}{64}\right)^{\frac{4}{3}}+4\times\left(\frac{27}{64}\right)-2=\frac{203}{128}$

$\left(\frac{27}{64},\frac{203}{128}\right)$

Links to Other Concepts

- Polynomials
- Algebraic manipulation
- Indices
- Mechanics, rates of change, velocity, acceleration
- Integration
- Problem solving
- Binomial expansion
- Maximisation/optimisation problems
- Growth and decay

SUMMARY

- $f'(x) = \lim_{\delta x \to 0} \frac{f(x+\delta x) - f(x)}{\delta x}$
- **To differentiate, multiply the coefficient by the power, then subtract 1 from the power.** $\frac{d}{dx} px^a = apx^{a-1}$
- **Use index notation to express roots and situations where the x term is in the denominator to get an expression into a differentiable form.**
- **Expand brackets and simplify before differentiating.**
- **To find a gradient at a given point: find the gradient function by differentiating y with respect to x $\left(\frac{dy}{dx}\right)$ then substitute the value of x for which you want to know the gradient. Be aware that there may be two solutions if there is a square root.**
- **If using a calculator to perform calculus, still show the working carried out.**

QUICK TEST

1. Differentiate each of the following:

   a) $3x^2$ b) $5x$ c) $x^{\frac{4}{5}}$ d) $\frac{3}{\sqrt{x}}$

   e) $\frac{4}{x^3}$ f) $3x+2$ g) $5x^4+3x^3-3$

2. $y=(2+\sqrt{x})^3$

   a) Expand the brackets and express in powers of x.

   b) Hence find $\frac{dy}{dx}$

   c) Find the possible gradients at the point when $x = 4$.

3. $y=\frac{(4+x^3)\sqrt[3]{x}}{x^{\frac{7}{3}}}$

   a) Find $\frac{dy}{dx}$

   b) Find the gradient at the point when $x = 2$.

PRACTICE QUESTIONS

1. $y=(3\sqrt{x}+2x^2)^3$

   a) Expand and simplify the expression leaving all terms in index form. **[4 marks]**

   b) Find $\frac{dy}{dx}$ **[2 marks]**

   c) Hence find the gradient when $x = 4$. **[2 marks]**

2. a) The graph shows the function $y=f(x)$. Estimate the gradient at $x = 3$. **[3 marks]**

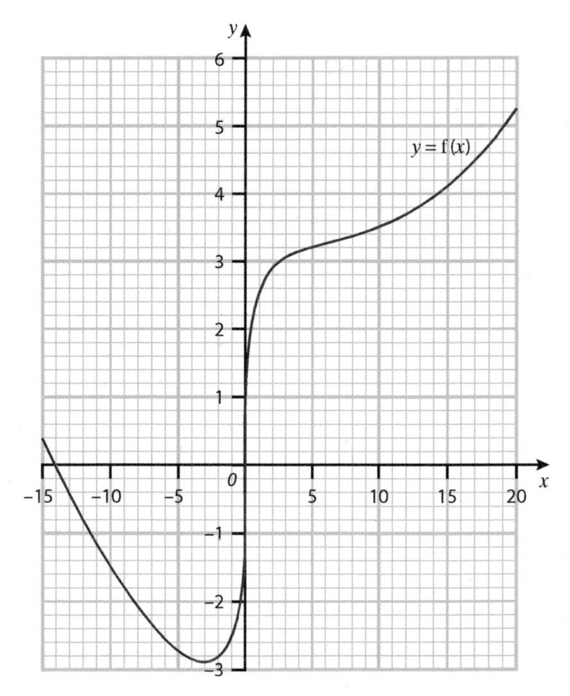

   b) This graph shows the function $y=g(x)$. Sketch the function $y=g'(x)$. **[3 marks]**

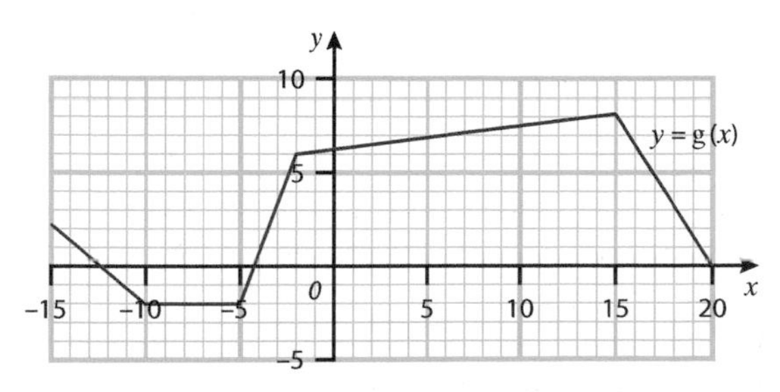

Differentiation 2

Increasing and Decreasing Functions

Identifying whether a function is increasing or decreasing can help with sketching graphs or determining the nature of what it represents within a context. To determine if a function is increasing or decreasing at a given point, find $\frac{dy}{dx}$ and substitute in the given x-value to find the gradient at that point:

- If $\frac{dy}{dx} > 0$, then the function is **increasing**.
- If $\frac{dy}{dx} < 0$, then the function is **decreasing**.
- If $\frac{dy}{dx} = 0$, then the function is **stationary**.

To change from an increasing function to a decreasing function, the graph must either have a stationary point (true for all continuous graphs) or pass an asymptote.

Example

Determine if the function $f(x) = 2x^3 + 3x^2 - 12x + 1$ is increasing, decreasing or stationary at points:

(1, –6) (–3, 10) (2, 5) (–1, 14)

The first step is to find $f'(x)$ in each case.

$f'(x) = 6x^2 + 6x - 12$

$f'(1) = (6 \times 1^2) + (6 \times 1) - 12 = 0$

At (1, –6) the graph is stationary.

$f'(-3) = 6 \times (-3)^2 + 6 \times (-3) - 12 = 24$

At (–3, 10) the graph is increasing.

$f'(2) = (6 \times 2^2) + (6 \times 2) - 12 = 24$

At (2, 5) the graph is increasing. Interesting to note it is increasing at the same rate as at (–3, 10).

$f'(-1) = 6 \times (-1)^2 + 6 \times (-1) - 12 = -12$

At (–1, 14) the graph is decreasing.

Finding Stationary Points

To find a stationary point, find the gradient function $\left(\frac{dy}{dx}\right)$ and set it equal to zero and solve for x. If asked to find the coordinates of the stationary point, substitute the value found for x back into the original equation to find the y-value.

Example

The function $f(x) = 2x^3 + 3x^2 - 12x + 1$ has one stationary point at (1, –6). Find the coordinates of the second stationary point.

$f'(x) = 6x^2 + 6x - 12$

$6x^2 + 6x - 12 = 0$

$6(x^2 + x - 2) = 0$

$x^2 + x - 2 = 0$

$(x - 1)(x + 2) = 0$

Having already found the stationary point at $x = 1$, the second point is at $x = -2$.

$f(-2) = 2(-2)^3 + 3(-2)^2 - 12(-2) + 1 = 21$

The second stationary point is at (–2, 21).

The Nature of Stationary Points

Stationary points, or turning points, are significant as they occur when the gradient of a curve momentarily equals zero.

To determine the nature of the turning point (i.e. whether it is a maximum, a minimum or a point of inflection), the **second differential** $\left(f''(x) \text{ or } \frac{d^2y}{dx^2}\right)$ can be taken.

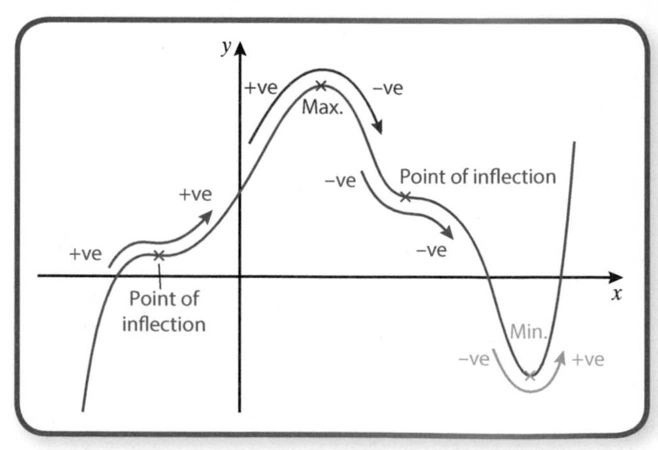

At a minimum the gradient is 0 as it changes from a negative to a positive gradient. This is implied by a positive second differential (the overall change from negative to positive is a positive change).

If $f'(x) = 0$ and $f''(x) > 0$, the stationary point is a minimum.

At a **maximum** the gradient is 0 as it changes from a positive to a negative gradient. This is implied by a negative second differential (the overall change from positive to negative is a negative change).

If $f'(x) = 0$ and $f''(x) < 0$, the stationary point is a maximum.

At a **point of inflection** there is a change in the curvature of the graph. It is a point when the second differential equals 0.

Note: Whilst a point of inflection will have a second differential equal to 0, the second differential being 0 is not enough to imply that it is a point of inflection. Points of inflection can have a first differential of 0.

Example

The function $f(x) = 2x^3 + 3x^2 - 12x + 1$ has stationary points at $(-2, 21)$ and $(1, -6)$. Determine the nature of each of these turning points.

First find $f''(x)$: $f'(x) = 6x^2 + 6x - 12$

$f''(x) = 12x + 6$

At $(-2, 21)$, $f''(x) = 12 \times (-2) + 6 = -18$

$-18 < 0$ $\therefore$ the stationary point is a maximum.

At $(1, -6)$, $f''(x) = 12 \times (1) + 6 = 18$

$18 > 0$ $\therefore$ the stationary point is a minimum.

Note: A minimum found this way could be the overall minimum value for the graph but could also be what is known as a local minimum (i.e. a point where the curve makes a U shape but there are other values of x that would give a smaller value for y).

The Meaning of Differentials and Second Differentials – Context

The point of differentiation is to find the rate of change (the gradient of the graph at a specific point). This can have different meanings depending on the context.

Example

The rate of change of displacement with respect to time is the velocity.

The rate of change of velocity with respect to time is acceleration.

The rate of change of volume of water in a tank with respect to time is m^3s^{-1} (volume per second).

The second differential finds the rate of change of the gradient: 'the rate of change of the rate of change'!

Example

The second derivative of displacement with respect to time is acceleration.

Tangents, Normals and Coordinates

By finding gradients of curves, you can find the gradient of the tangent to the curve at any given point, and hence find the gradient of the normal to the curve.

Finding the equation of a straight line: $y_2 - y_1 = m(x_2 - x_1)$.

To use this equation, you need:

1. A point (x_1, y_1) that the line passes through.
2. Either a second point (which can be used to find the gradient) or the gradient of the line (m).

Tangent – a line that has the same gradient as the curve at that specific point. If asked to find the equation of the tangent to a curve, start by differentiating and substitute in the given x-value (if an x-value isn't given, you first need to find it).

Normal – a line that is perpendicular to the tangent. It has a gradient that is the **negative reciprocal** of the gradient of the curve. If the curve has a gradient of a, then the gradient of the normal is $-\frac{1}{a}$.

Example

a) Find the equation of the tangent to the curve $y = 3x^3 - 12x$ at the point where $x = -1$.

Find the y-coordinate: $y = 3(-1)^3 - 12(-1)$

$= -3 + 12 = 9$

Curve passes through point $(-1, 9)$.

Find the gradient of the tangent:

$\frac{dy}{dx} = 9x^2 - 12$

At $x = -1$, $\frac{dy}{dx} = 9 - 12 = -3$

The gradient of the tangent $m = -3$.

$y - y_1 = m(x - x_1)$
$y - 9 = -3(x - -1)$
$y - 9 = -3x - 3$

As no particular form is required by the question, use either $ax + by + c = 0$ or $y = mx + c$:

$y = -3x + 6$

b) Find the coordinates of the point where the tangent intercepts the curve.

To find intersections, set the equations equal to each other:

$-3x + 6 = 3x^3 - 12x$
$3x^3 - 9x - 6 = 0$
$x^3 - 3x - 2 = 0$

This gives a cubic so there will be three roots to the equation. Sketching the original graphs will help:

$y = 3x^3 - 12x = x(3x^2 - 12) = x(3x - 6)(x + 2)$,
so positive cubic curve with roots at $x = -2$, $x = 0$ and $x = 2$.

Tangent at (−1, 9) has negative gradient so is to the right of the local maximum.

Line $y = -3x + 6$ has x-intercept at 2.

The answer to the equation $x^3 - 3x - 2 = 0$ will have a repeat root at (−1, 9), as it touches but does not cross the line here. The other root is at (2, 0), by inspection of graphs.

Check these are **all** the roots to the equation $x^3 - 3x - 2 = 0$ by expansion (alternate methods could be used):

$(x + 1)^2 (x - 2) = (x^2 + 2x + 1)(x - 2)$
$= x^3 + 2x^2 - 2x^2 + x - 4x - 2$
$= x^3 - 3x - 2$

The point where the tangent crosses the curve is at $x = 2$

$y = (3 \times 2^3) - (12 \times 2) = 0$

The point is (2, 0).

c) Find the exact greatest distance between any two points where the normal intersects the curve.

The gradient of the normal is $-\frac{1}{-3} = \frac{1}{3}$ and it passes through the point (−1, 9).

Equation of the normal $y - y_1 = m(x - x_1)$

$y - 9 = \frac{1}{3}(x - -1)$
$3y - 27 = x + 1$
$3y = x + 28$
$y = \frac{1}{3}x + \frac{28}{3}$

Find points of intersection with the curve by setting them equal to each other:

$3x^3 - 12x = \frac{1}{3}x + \frac{28}{3}$
$3(3x^3 - 12x) = x + 28$
$9x^3 - 36x = x + 28$
$9x^3 - 37x - 28 = 0$

In this case you know one solution already, $x = -1$, which means that $(x + 1)$ is a factor.

$$\begin{array}{r|l} & 9x^2 - 9x - 28 \\ \hline x+1 & 9x^3 + 0x^2 - 37x - 28 \\ & \underline{9x^3 + 9x^2} \\ & \quad -9x^2 - 37x \\ & \quad \underline{-9x^2 - 9x} \\ & \qquad -28x - 28 \\ & \qquad \underline{-28x - 28} \end{array}$$

The other two intersections will be found by solving $9x^2 - 9x - 28 = 0$

$x = \frac{-b \pm \sqrt{b^2 - 4ac}}{2a}$

$x = \frac{9 \pm \sqrt{(-9)^2 - 4 \times 9 \times -28}}{2 \times 9}$

$x = \frac{7}{3} \rightarrow y = \frac{1}{3} \times \frac{7}{3} + \frac{28}{3} = \frac{91}{9}$

$x = -\frac{4}{3} \rightarrow y = \frac{1}{3} \times \frac{-4}{3} + \frac{28}{3} = \frac{80}{9}$

From the previous diagram, it is possible to see that the greatest distance is between the x-coordinates $-\frac{4}{3}$ and $\frac{7}{3}$.

Find distance using Pythagoras' Theorem:

Distance $= \sqrt{\left(\frac{11}{3}\right)^2 + \left(\frac{11}{9}\right)^2} = \frac{11\sqrt{10}}{9}$

SUMMARY

- To differentiate, multiply the coefficient by the power, then subtract 1 from the power. $\frac{d}{dx}px^a = apx^{a-1}$
- If $\frac{dy}{dx} > 0$, then the function is increasing.
- If $\frac{dy}{dx} < 0$, then the function is decreasing.
- If $\frac{dy}{dx} = 0$, then the function is stationary.
- To find the coordinates of a stationary point:
  - differentiate, set equal to zero and solve
  - find the y-coordinate by substituting the x-value(s) into the original equation.
- $f''(x)$ or $\frac{d^2y}{dx^2}$ means the second differential. It is the rate of change of the gradient.
- To find a second differential of a given function, differentiate the function twice.
- If $f'(x)=0$ and $f''(x)>0$, the stationary point is a minimum.
- If $f'(x)=0$ and $f''(x)<0$, the stationary point is a maximum.
- Tangents have the same gradient as the function at the given point. This can be found using differentiation.
- Normals have the negative reciprocal of the gradient of the tangent.
- To find a point where a curve is intersected by a normal or a tangent, one result is already given and so can be divided out.

Links to Other Concepts

- Coordinate geometry
- Polynomials
- Algebraic manipulation
- Indices
- Mechanics, rates of change, velocity, acceleration
- Integration
- Growth and decay
- Quadratics, factor theorem

QUICK TEST

1. Determine if these functions are increasing, decreasing or stationary at the point stated.
   a) $y = 2x^2 - 4x + 1$, at $x = 2$
   b) $y = -2x^2 - 4x + 1$, at $x = -1$
   c) $y = x^{0.5} - x$, at $x = 1$
   d) $y = x^{\frac{4}{3}} + x^3$, at $x = 0$
   e) $y = x(4x+5)(x+2)$, at $x = -1$
2. Find the second differential of these:
   a) $y = x^2$ b) $y = 2x^3 - 2x^2 + x$
3. Determine the location and nature of each of the stationary points of the curve $y = -x^3 + 3x^2 - 4$.
4. a) Find the gradient of the tangent to the curve $y = 3x^4 + 4x^3 - 3x$ at the point where $x = -1$.
   b) A line running parallel to the tangent has the equation $y = ax + 2$. Find the value of a.
5. Find the gradient of the normal to the function $f(x) = \frac{1}{3}x^{\frac{3}{2}} + 2x^{-1} - 2$ at $x = 4$.

PRACTICE QUESTIONS

1. The function $f(x) = \frac{2}{3}x^3 + 4x^2 + px - 2$ has a single, repeat, stationary point Q.
   a) Show that p must satisfy the equation $64 - 8p = 0$. **[4 marks]**
   b) Hence find the coordinates of the stationary point on the curve. **[4 marks]**
   c) If instead $p = 6$, find the coordinates for the two stationary points and state whether each is a maximum or a minimum. **[6 marks]**
2. The normal to the curve $y = 4x^3 + 6x^2 + x$ at point P, where $x = -1$, intersects the graph in two other places, Q and R.
   a) Find the coordinates of the points P, Q and R. **[9 marks]**
   b) Is the function increasing, decreasing or stationary at the point Q (which lies in-between P and R)? **[1 mark]**

Integration 1

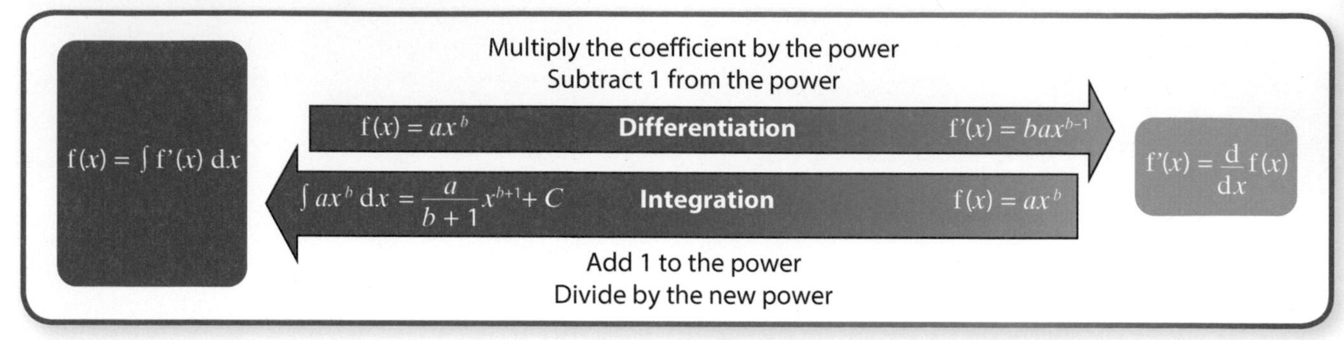

Integration

Integration is the inverse of differentiation. It can be used to find a function having been given the gradient function. It also can be used to find the area under a curve.

A constant (+ C) has to be added during integration, as for every gradient function there is a family of graphs that could represent the original function, all of them vertical translations of each other. The original function $f(x)$ may therefore have contained a constant which differentiated to zero and so 'disappeared' in the gradient function, so a '+ C' needs to be added whenever you integrate.

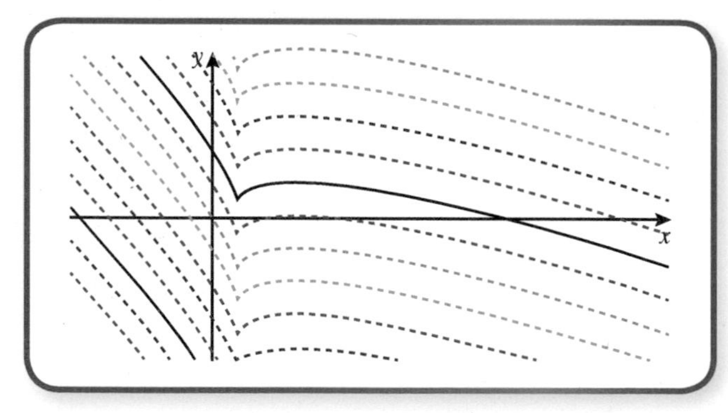

This diagram shows 'a family' of graphs $y = f(x) + a$. Each graph has the same gradient at corresponding values of x because each line is a vertical translation of the others. The function $f'(x)$ (the derivative) would be the same for each line whatever the value of a. Based only on $f'(x)$, it is impossible to know which of the lines, or the infinite other lines in the 'family', represents $f(x)$.

Example

Positive integer power:

$$\int 3x^2\,dx = \frac{3}{2+1}x^{2+1} + C = x^3 + C$$

Power of 1: $\int -4x\,dx = \frac{-4}{1+1}x^{1+1} + C = -2x^2 + C$

Power of 0 (or constant terms):

$$\int 2\,dx = \frac{2}{0+1}x^{0+1} + C = 2x + C$$

Negative integer power:

$$\int 6x^{-2}\,dx = \frac{6}{-2+1}x^{-2+1} + C = -6x^{-1} + C$$

Fractional power:

$$\int 3x^{\frac{5}{2}}\,dx = \frac{3}{\frac{5}{2}+1}x^{\frac{5}{2}+1} + C = \frac{3}{\frac{7}{2}}x^{\frac{7}{2}} + C = \frac{6}{7}x^{\frac{7}{2}} + C$$

$$\int x^n\,dx = \frac{x^{n+1}}{n+1} + C, \quad n \neq -1$$

Note: The power will not equal −1 at AS-level.

A note on notation: $\int f'(x)\,dx$ means the integral of $f'(x)$ with respect to x. The dx is part of the integral function. Much like how brackets come in pairs, the integral sign must be followed by a dx. It is possible that a question might use different letters, or ask for a function to be integrated with respect to y, in which case dy would be used.

As with differentiation, the terms of an expression can be integrated one at a time. Only one $+C$ is needed, not one for each term. The next example shows more steps in the working than is strictly necessary. However, it is very easy to make mistakes with fractional and negative powers and including extra steps in your working can be a good way to avoid them.

Example

Integrate the function $y=12x^4+3x^{-\frac{3}{4}}-3$.

$$\int y\,\mathrm{d}x=\int 12x^4+3x^{-\frac{3}{4}}-3\,\mathrm{d}x$$

$$=\frac{12}{4+1}x^{4+1}+\frac{3}{-\frac{3}{4}+1}x^{-\frac{3}{4}+1}-\frac{3}{(0+1)}x^{0+1}+C$$

$$=\frac{12}{5}x^5+12x^{\frac{1}{4}}-3x+C$$

$$\int ax^n+bx^m\mathrm{d}x=a\int x^n\,\mathrm{d}x+b\int x^m\,\mathrm{d}x+C,$$

$$n\neq-1,\ m\neq-1$$

Finding C

As all constants are reduced to zero by differentiation, it isn't possible to know the value of C just by integration. To find C, extra information is needed. This is generally in the form of a point on the graph $y=\mathrm{f}(x)$.

Example

$\mathrm{f}'(x)=3x^2-4x+2$. Find $\mathrm{f}(x)$ given that when $x=3$, $\mathrm{f}(x)=7$.

$$\mathrm{f}(x)=\int 3x^2-4x+2\,\mathrm{d}x$$

$$=\frac{3}{3}x^3-\frac{4}{2}x^2+2x+C=x^3-2x^2+2x+C$$

Substitute in given values and solve for C:

$7=3^3-(2\times3^2)+(2\times3)+C$

$7=27-18+6+C$

$7=15+C$

$C=-8$

$\mathrm{f}(x)=x^3-2x^2+2x-8$

Always check the form of the final answer. In this example the question has asked for $\mathrm{f}(x)$.

Converting Expressions

As with differentiation, one of the biggest challenges can be getting the expression into a form which can be integrated to start with. Consider expanding brackets and converting roots and x terms in denominators using index form (see pages 4–5).

Example

$$\frac{\mathrm{d}y}{\mathrm{d}x}=\left(1+\frac{1}{x\sqrt{x}}\right)^3$$

Find an expression for y in terms of x.

$$\left(1+\frac{1}{x\sqrt{x}}\right)^3=\left(1+x^{-\frac{3}{2}}\right)^3$$

$$=1+3\times x^{-\frac{3}{2}}+3\times\left(x^{-\frac{3}{2}}\right)^2+\left(x^{-\frac{3}{2}}\right)^3$$

$$=1+3x^{-\frac{3}{2}}+3x^{-3}+x^{-\frac{9}{2}}$$

$$y=\int\left(1+\frac{1}{\sqrt{x}}\right)^3\mathrm{d}x$$

$$=\int 1\,\mathrm{d}x+3\int x^{-\frac{3}{2}}\,\mathrm{d}x+3\int x^{-3}\,\mathrm{d}x+\int x^{-\frac{9}{2}}\,\mathrm{d}x$$

$$y=x-6x^{-\frac{1}{2}}-\frac{3}{2}x^{-2}-\frac{2}{7}x^{-\frac{7}{2}}+C$$

Algebraic Constants

As for differentiation, much of the calculation can be done using a graphical calculator. One way to check understanding of the process is to use algebraic constants in the question. An algebraic constant just represents a number, so the method is exactly the same but skills in algebraic manipulation may be relevant. Equating coefficients is a skill that may well be useful, along with simultaneous equations, to find the values of the unknowns.

It is worth remembering that, if the curve passes through a point $(0, a)$, the constant of integration $C=a$. If a curve passes through the origin, there is no constant term in its equation.

DAY 4

Example

$f'(x)=3x^a+2x^2, a\neq-1$

a) Find an expression for $f(x)$.

$f(x)=\frac{3}{(a+1)}x^{a+1}+\frac{2}{3}x^3+C$

b) The curve passes through the points (1, 2) and $\left(0,-\frac{1}{6}\right)$.

Find the values of a and C.

$f(0)=-\frac{1}{6}=\frac{3}{(a+1)}\times 0^{a+1}+\frac{2}{3}\times 0^3+C=C$

$C=-\frac{1}{6}$

$f(1)=2=\frac{3}{(a+1)}\times 1^{a+1}+\frac{2}{3}\times 1^3-\frac{1}{6}$

$\frac{3}{(a+1)}+\frac{2}{3}-\frac{1}{6}=2$

$\frac{3}{a+1}=2+\frac{1}{6}-\frac{2}{3}=\frac{3}{2}$

$a+1=2$

$a=1$

SUMMARY

- $\int x^n\,dx=\frac{x^{n+1}}{n+1}+C, n\neq-1$
- **To integrate, add 1 to the power then divide the coefficient by the new power.**
- $\int ax^n+bx^m\,dx=a\int x^n\,dx+b\int x^m\,dx+C,$ $n\neq-1, m\neq-1$
- **The $+C$ needs to be added to define the curve fully. Don't forget it!**
- **Use given values for the curve in order to find C.**
- **Equations may need to be manipulated into a form that can be integrated. This might involve expanding brackets, binomial expansion, simplifying expressions or use of index notation.**

Links to Other Concepts

- Polynomials
- Algebraic manipulation
- Indices
- Mechanics, velocity, acceleration
- Differentiation
- Problem solving
- Maximisation/optimisation problems
- Binomial expansion
- Rates of change

QUICK TEST

1. Integrate each of the following:

   a) $3x^2$ b) $5x$

   c) $x^{\frac{4}{5}}$ d) $\frac{3}{\sqrt{x}}$

   e) $\frac{4}{x^3}$ f) $3x+2$

   g) $5x^4+3x^3-3$

2. $f'(x)=7x^{\frac{4}{3}}-2x^{-2}$

   a) Find $f(x)$.

   b) Given that $f(8) = 400$, find C.

3. The acceleration of a car is given by the equation $a = 3t + 2$.

   a) Find the velocity of the car by integrating with respect to time.

   b) Given that at time $t = 0$ the velocity of the car is $10\,\text{ms}^{-1}$, find C.

PRACTICE QUESTIONS

1. The gradient function $\frac{dy}{dx}=10x+2$, when plotted on the same axes as the quadratic curve, forms a tangent to the curve at the point $x = 0.8$.

   a) Show that both the gradient function and the quadratic curve pass through the point $(0.8, 10)$. **[2 marks]**

   b) Find the equation of the quadratic curve. **[4 marks]**

2. $f'(x)=ax^2+2x+a^2$, where a is a constant and $a>0$.

   a) Find an expression for $f(x)$. **[2 marks]**

   b) Given that the function $f(x)$ passes through the coordinates $(0, -7)$ and $(1, -4.25)$, find the value of a. **[6 marks]**

3. $y=(p\sqrt{x}+2x)(x^q-1)$, where p and q are positive constants.

   a) Show that $y=px^{q+\frac{1}{2}}+2x^{q+1}-px^{\frac{1}{2}}-2x$. **[2 marks]**

   b) Hence find an expression, in terms of p and q, for the integral of y. **[4 marks]**

   c) Given that $\int y\,dx=4x^3+\frac{4}{7}x^{\frac{7}{2}}-8x^{\frac{3}{2}}-x^2+C$, find the values of p and q. **[3 marks]**

Integration 2

From the last topic:

$$\int x^n \mathrm{d}x = \frac{x^{n+1}}{n+1} + C, n \neq -1$$

Rates of Change and the Meaning of Integration

Differentiating is finding the rate of change, which on a graph is represented by the gradient of the curve. Integrating finds the area between the function and the x-axis.

Between Bounds

The area to be found could be specified as between bounds. In the graph shown, the area is between the x-axis, the curve and the lines $x = 1$ and $x = 4$. A small number is used at the top and bottom of the integration symbol to define the bounds.

$A = \int_a^b \mathrm{f}(x)\mathrm{d}x$ means the integral between a and b of the function $\mathrm{f}(x)$ with respect to x.

Example

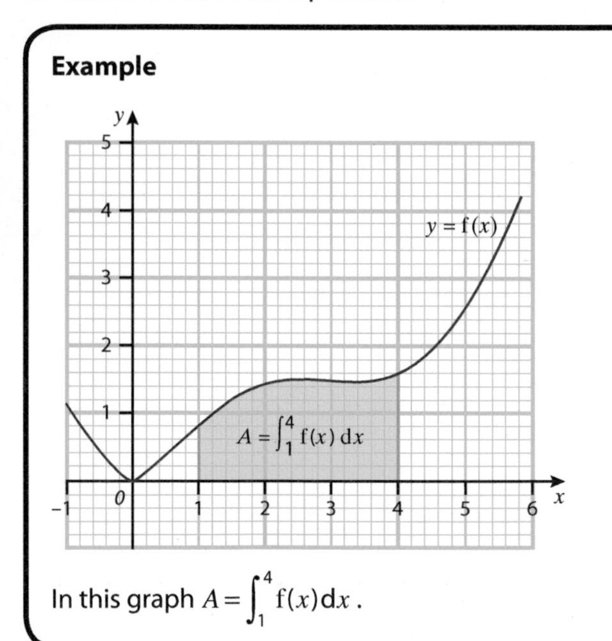

In this graph $A = \int_1^4 \mathrm{f}(x)\mathrm{d}x$.

When finding an integral between bounds, it is important to show the following steps.

Example

$\mathrm{f}(x) = x(x - 4)$. Find the area between the curve, the x-axis and the lines $x = 3$ and $x = 1$.

- Rearrange $\mathrm{f}(x)$ into a form that is easily integrated if necessary (simplified index form):

  $\mathrm{f}(x) = x(x - 4) = x^2 - 4x$

- Consider sketching the curve if it could be helpful. Sketches are generally good for visualising what is being done and helping to spot unreasonable answers, but don't waste time trying to sketch a curve if it's not necessary. Sometimes it may be asked for as a step in the question. At other times it might be necessary to support your working even if not specified.

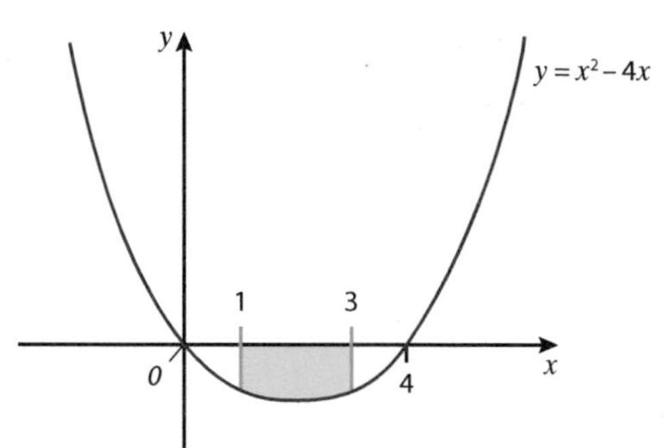

- State the integration to be performed with limits:

  $A = \int_a^b \mathrm{f}(x)\ \mathrm{d}x$

  $A = \int_1^3 x^2 - 4x\ \mathrm{d}x$

- Use square brackets to show the integrated function:

  $A = \left[\frac{1}{3}x^3 - 2x^2\right]_1^3$

- Show the method for finding the area between bounds, which is $(\mathrm{g}(b)) - (\mathrm{g}(a))$, i.e. substitute the bounds into the integrated function and subtract the lower bound from the upper:

  $A = \left(\frac{1}{3}(3)^3 - 2(3)^2\right) - \left(\frac{1}{3}(1)^3 - 2(1)^2\right)$

- Solve (or simplify if algebraic) to get a numerical answer:
  $A=(-9)-\left(-\frac{5}{3}\right)=-\frac{22}{3}$
- Check the answer is reasonable and convert it to positive if it is negative (because it is an area below the x-axis). If the question is context-based, convert it back into context and include units if relevant.
  $A=\frac{22}{3}$ units^2
  Note: A negative result was expected as the area shown on the sketch was below the x-axis.

Don't forget

The integrated function **doesn't need to include $+C$ when between bounds**, as it would cancel out anyway when the two values are subtracted from each other:

$$A=\left[g(x)+C\right]_a^b$$
$$=(g(b)+C)-(g(a)+C)$$
$$=g(b)+C-g(a)-C$$
$$=g(b)-g(a)$$

Including a $+C$ isn't wrong, but it can add a small extra complication which can be avoided easily. Just remember the $+C$ when not between bounds.

Between Roots

The question might not specify the bounds but instead require the roots to be found. In this case the same process is followed but with a first step of identifying the bounds by solving the equation. Note: If the area is bounded by the curve, the positive x-axis and the positive y-axis, one of the bounds is $x=0$ and the other is a root to the equation to be found.

Example

Find the area enclosed by the curve $y=3x^2-5x-2$ and the x-axis.

The first step is to set the equation equal to zero and solve to find roots. Sketch the curve to show the roots and the shape to help identify the area to be found:

$3x^2-5x-2=0$

$(3x+1)(x-2)=0$

$x=-\frac{1}{3}$ and $x=2$

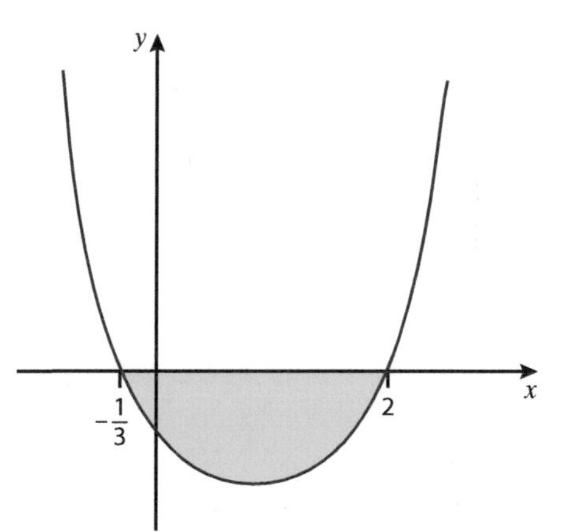

Now following the steps shown previously:

$$A=\int_{-\frac{1}{3}}^{2}(3x^2-5x-2)\,dx$$
$$=\left[x^3-\tfrac{5}{2}x^2-2x\right]_{-\frac{1}{3}}^{2}$$
$$=\left(2^3-\tfrac{5}{2}\times 2^2-2\times 2\right)-\left(\left(-\tfrac{1}{3}\right)^3-\tfrac{5}{2}\times\left(-\tfrac{1}{3}\right)^2-2\times\left(-\tfrac{1}{3}\right)\right)$$
$$=-6\frac{19}{54}=-\frac{343}{54}$$

Area $=\frac{343}{54}$ units^2

Composite Areas

At GCSE, you had to identify the component shapes in order to find the areas of composite shapes. The composite areas were added or subtracted to find the required area.

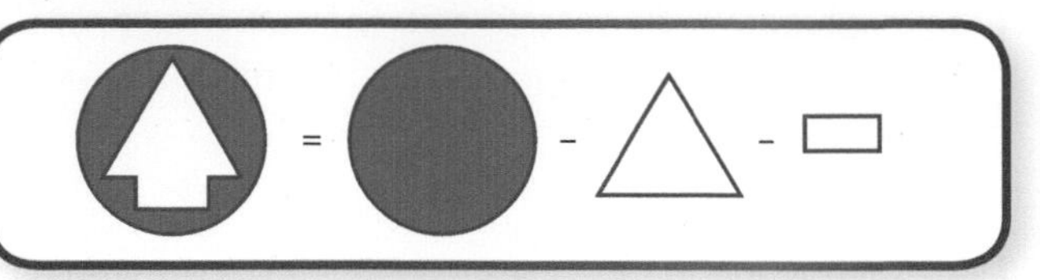

Exactly the same skills are needed to find the composite areas formed between straight lines and curves.

Shapes useful to know:

- Area of a right-angled triangle $= \frac{1}{2}bh$
- Area of a trapezium $= \frac{1}{2}(a+b)h$
- Area of a rectangle $= bh$

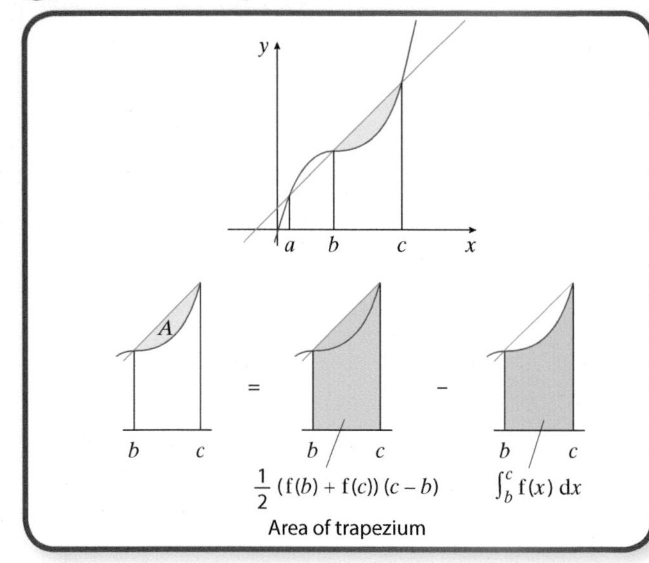

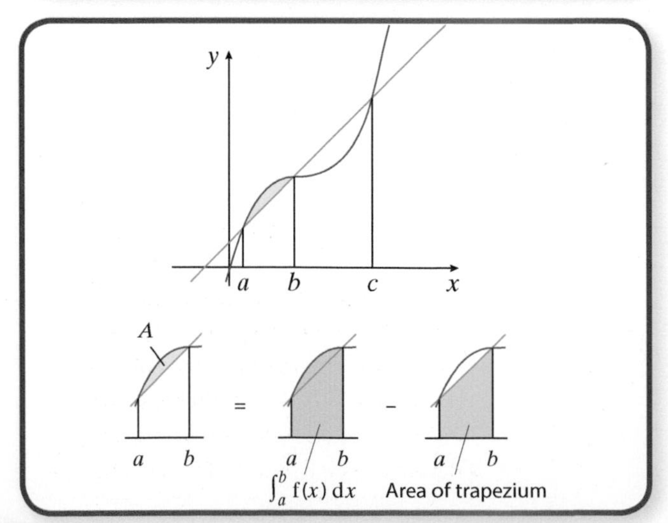

Consider the areas that need to be subtracted or added to find the area. Find each individually then combine to find the area required.

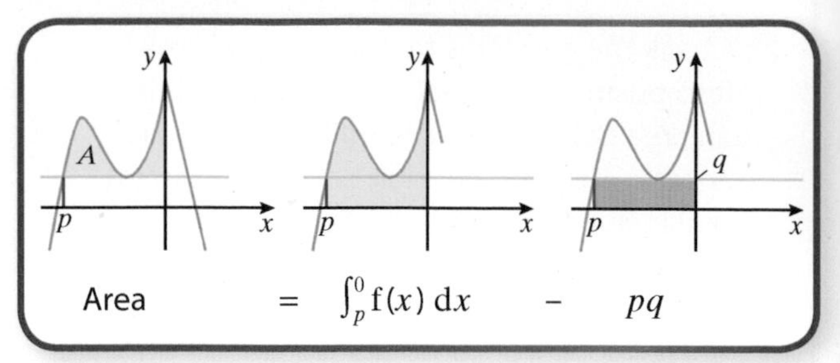

Example

The graph shows the curve $y = 3x^5 + x^3 + 2$ and the straight line $y = 4x + 2$.

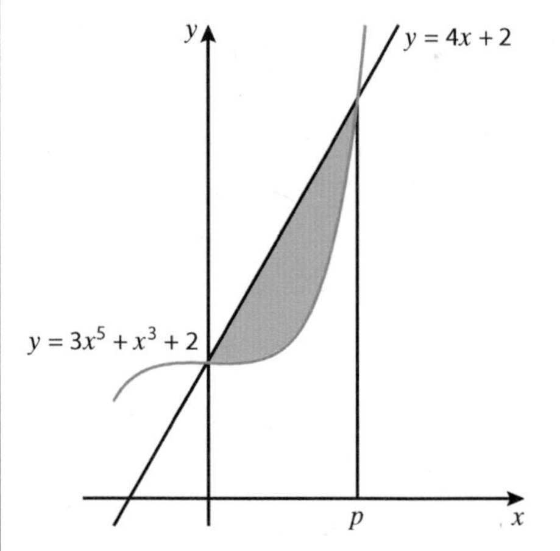

a) Find the value of p and the coordinates of the intersection at $x = p$.

$$3x^5 + x^3 + 2 = 4x + 2$$
$$3x^5 + x^3 - 4x = 0$$
$$x(3x^2 + 4)(x^2 - 1) = 0$$
$$x = 0$$
$3x^2 + 4 = 0 \rightarrow$ No real results

$x^2 - 1 = 0 \quad \rightarrow x = 1, x = -1$

$x = 0,\ x = -1,\ x = 1$

As p is positive, as shown on graph, $p = 1$.

At $x = 1 \qquad y = (4 \times 1) + 2 = 6$

Point of intersection is $(1, 6)$.

b) Find $\int_0^p 3x^5 + x^3 + 2 \, dx$.

$$\int_0^1 3x^5 + x^3 + 2 \, dx = \left[\tfrac{1}{2}x^6 + \tfrac{1}{4}x^4 + 2x\right]_0^1$$
$$= \left(\tfrac{1}{2} + \tfrac{1}{4} + 2\right) - (0)$$
$$= \tfrac{11}{4}$$

c) Hence find the area of the shaded region.

Area of shaded region = area of trapezium – area under curve

To find the area of the trapezium, work out the values of a, b and h (see diagram):

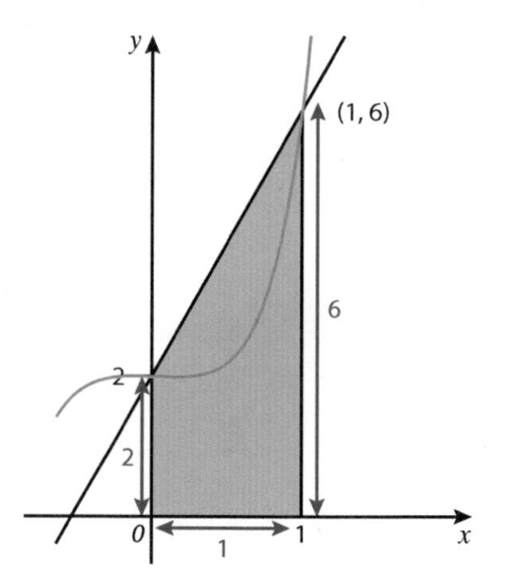

$$A = \tfrac{1}{2}(2+6) \times 1 = 4$$

$$\text{Shaded area} = 4 - \tfrac{11}{4}$$
$$= \tfrac{5}{4} = 1.25 \text{ units}^2$$

Context

There are strong links to mechanics topics. In context, questions might not mention area at all but instead talk about displacement having given a function defining the velocity in terms of time (velocity–time graph, where the area under the curve is the displacement). The modulus of displacement is distance travelled. In this case the positive value is taken. If asked for displacement, keep the negative value if there is one.

Example

Find the distance travelled by a car moving with velocity $v = t^2 - 6t + 10$ ms^{-1}, between times $t = 2$ s and $t = 6$ s.

$$\int_2^6 t^2 - 6t + 10 \, dx = \left[\tfrac{1}{3}t^3 - 3t^2 + 10t\right]_2^6$$
$$= \left(\tfrac{1}{3} \times 6^3 - 3 \times 6^2 + 10 \times 6\right) - \left(\tfrac{1}{3} \times 2^3 - 3 \times 2^2 + 10 \times 2\right)$$
$$= 24 - \tfrac{32}{3} = \tfrac{40}{3}$$

Distance travelled between 2 and 6 seconds = 13.33 metres (2 d.p.).

Common Mistakes

The common mistakes for integrating with bounds are the same as for most algebraic topics:

- Miscalculating with negatives.
- Keying everything into a calculator and not showing working – using a calculator is a good idea but make sure you show on paper the steps that the calculator is doing for you.
- Confusing equations. When using calculus, there are at least two functions generated. If finding points of intersection, that creates yet another function of x. It is possible to confuse which equation is the one to use. Showing clear working and including a few words can help to show that 'this' is the curve, 'this' is the integrated function, 'this' is the equation to solve for intercepts. If substituting back into an equation to find the y-coordinate, having solved the points where a line crosses the function, the linear equation is generally the easier to get an answer from (but using both would give a good way of checking the answer if there is time).

Links to Other Concepts

- Graph sketching
- Polynomials
- Algebraic manipulation
- Indices
- Mechanics, velocity, acceleration
- Differentiation
- Problem solving
- Maximisation/optimisation problems
- Binomial expansion
- Rates of change

SUMMARY

- $\int x^n \, dx = \frac{x^{n+1}}{n+1} + C,\ n \neq -1$
- **Integration can be used to find the area between a curve and the x-axis.**
- $A = \int_a^b f(x)\,dx$ **means the integral between b and a of the function $f(x)$ with respect to x. It finds the area between the curve $y = f(x)$ and the x-axis.**
- **If the area is below the x-axis, the integral between bounds will return a negative answer. Note: At AS-level the area will be fully on one side or the other of the x-axis.**
- **If the integration between limits gives a negative answer, the area is below the x-axis. If asked for the area, take the modulus (i.e. make it positive).**
- **Composite areas could be made up using a curve and a straight line.**
- **For a composite area, one part will be the integral between limits and the other will be:**
  - **area of a right-angled triangle** $= \frac{1}{2}bh$
  - **area of a trapezium** $= \frac{1}{2}(a+b)h$
  - **area of a rectangle** $= bh$
- **Think about context and make sure your answer relates to the question posed.**

QUICK TEST

1. Find the area of the following shapes.

   a)
   6
   4

   b) 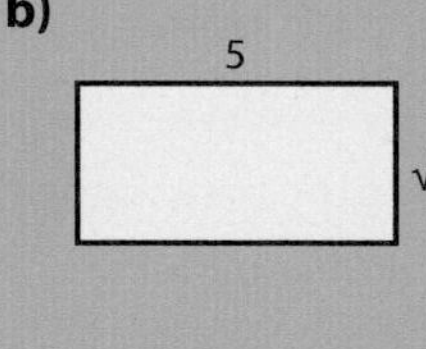

   c) 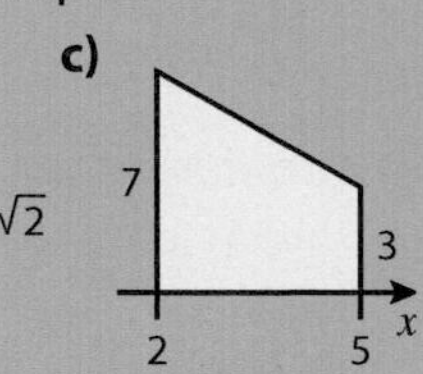

2. Will the answers to the following integrals be positive or negative?

   a) 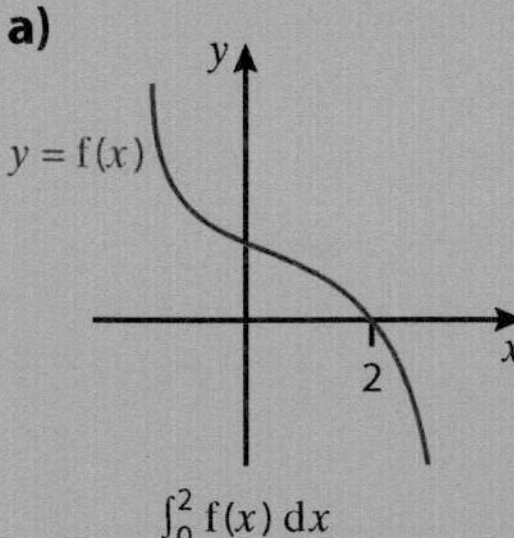

   $\int_0^2 f(x)\,dx$

   b) 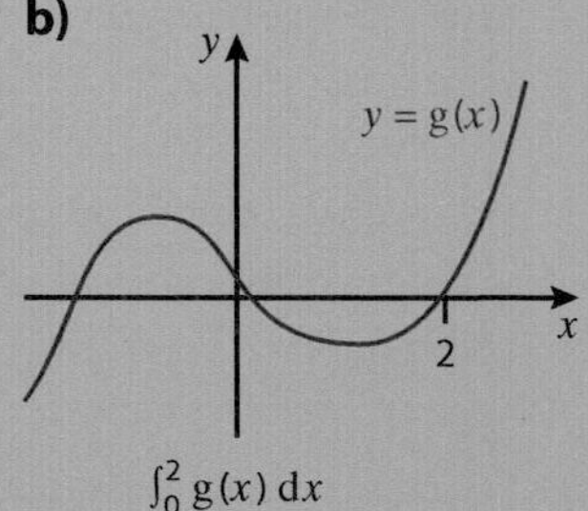

   $\int_0^2 g(x)\,dx$

3. Find the value of $\left[x^2 + x\right]_2^5$

4. Evaluate $\int_{-2}^{4} 6x^2\,dx$.

5. Find the area bounded by the line $y = -2x$, the x-axis and the lines $x = 2$ and $x = 7$.

6. The sketch shows the curve of $y = 2x^3 - 4x + 7$.

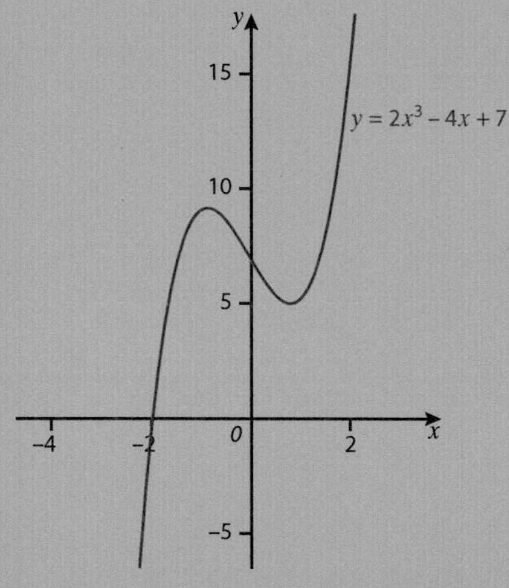

The working below shows how to find the area bounded by the curve, the negative x-axis and the positive y-axis, but some values have been blanked out. Copy and complete the working.

$$\int_{◆}^{◆} 2x^3 - 4x + 7\,dx = \left[\tfrac{1}{2}x^{◆} - ◆x^2 + 7x\right]_{◆}^{◆}$$

$$= \left(\tfrac{1}{2} \times ◆^{◆} - ◆ \times ◆^2 + 7 \times ◆\right)$$

$$- \left(\tfrac{1}{2} \times ◆^{◆} - ◆ \times ◆^2 + 7 \times ◆\right)$$

$$= (◆) - (◆) = ◆$$

PRACTICE QUESTIONS

1. Find the area bounded by the curve $y = -2x^2 + x + 6$ and the x-axis. **[5 marks]**

2. The graph shows the area enclosed by the curve $y = x^3 + 12x^{\frac{4}{3}} + 6x + 6$, the x-axis and the lines $x = 1$ and $x = -3$.

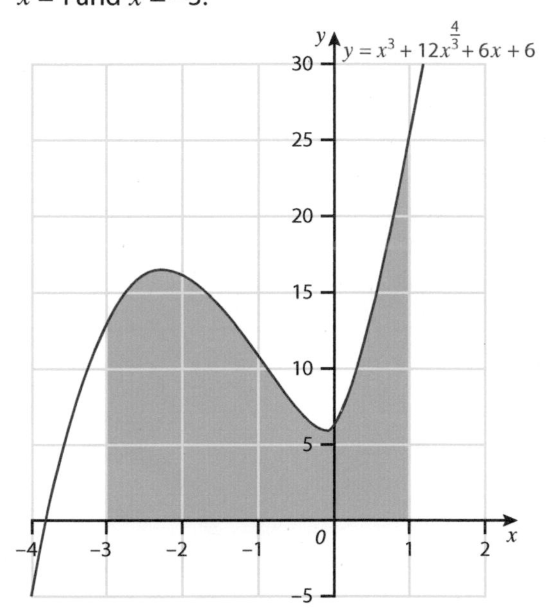

a) Find the area as shaded on the diagram. **[4 marks]**

b) Hence find the area bounded by the curve, the line $y = 25$, and the lines $x = 1$ and $x = -3$. **[3 marks]**

The line $y = 7x + 18$ intersects the curve at points P and Q as shown on the diagram.

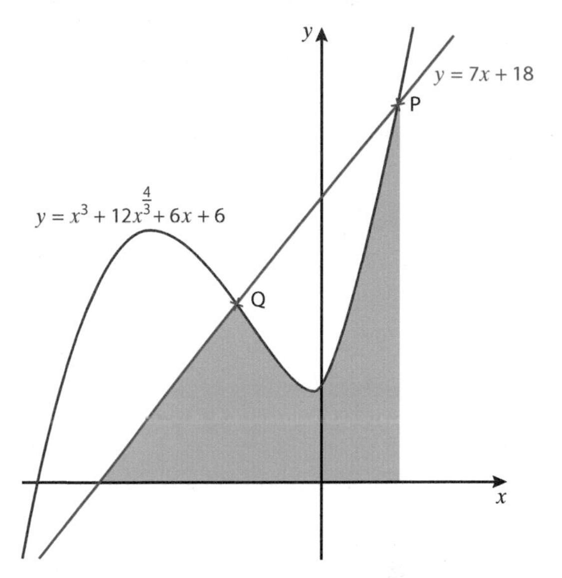

c) Show that the x-coordinates of P and Q must satisfy the equation $x^3 + 12x^{\frac{4}{3}} - x - 12 = 0$. **[2 marks]**

d) Sarah says the x-coordinates of P and Q are 1 and –1 respectively. Confirm, using the factor theorem, that Sarah is correct and hence find the coordinates of P and Q. **[2 marks]**

e) Hence find the area shaded on the graph. **[6 marks]**

Modelling, Quantities and Units

Mechanics is one of two applied sections within the AS syllabus (alongside Statistics). Mechanics is the study of the physical behaviour of objects. Applied units take skills and techniques learnt and use them to model real-life situations. Modelling starts simply and then is refined by comparing it with data taken in experiments.

The Fundamental Units

All measures of the physical motion of an object are based on the fundamental quantities of length, time and mass. Below are the details of the fundamental units of measure. You will also find the most common conversions. It is possible that other conversions may be needed. The converted units allow very small or very large measures to be represented simply. The SI (Système Internationale) is designed so that scientists, engineers and mathematicians can communicate clearly and concisely.

Length

Length measures a one-dimensional distance between two points. The SI base unit for length is **metres** (m).

Distance is a scalar quantity. Displacement is a vector quantity and, as such, has a direction associated with it both when it is positive and negative.

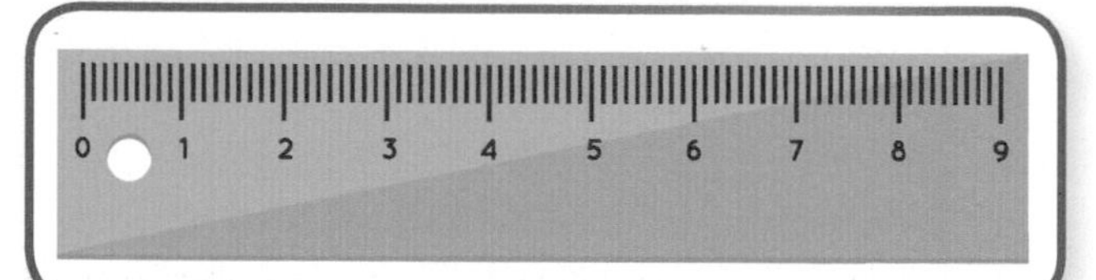

Converting Units of Length

1 km (kilometre) = 1000 m	To convert from kilometres to metres, multiply by 1000. To convert from metres to kilometres, divide by 1000.
1 cm (centimetre) = 0.01 m	To convert from centimetres to metres, multiply by 0.01 (or divide by 100). To convert from metres to centimetres, divide by 0.01 (which is equivalent to multiplying by 100).
1 mm (millimetre) = 0.001 m	To convert from millimetres to metres, multiply by 0.001 (or divide by 1000). To convert from metres to millimetres, divide by 0.001 (which is equivalent to multiplying by 1000).

Example
Ian swims at the local pool. The pool is 25 m long. He swims 102 lengths. How far is this in kilometres?

Distance swum $= 25 \times 102 = 2550$ m

Convert to km by dividing by 1000:

$\frac{2550}{1000} = 2.55$ km

Time

Time is measured using the SI base unit **seconds** (s).

Converting Units of Time

1 minute = 60 s	To convert from minutes to seconds, multiply by 60. To convert from seconds to minutes, divide by 60.
1 hour = 60 minutes = 3600 s	To convert from hours to seconds, multiply by 3600. To convert from seconds to hours, divide by 3600.

Often times are given in a mixture of units. Convert each part to the required unit and add them together.

Example
Express 2 hours, 14 minutes and 25 seconds in seconds.

25 s (already in seconds)

14 minutes $\times 60 = 840$ s

2 hours $\times 3600 = 7200$ s

Total time $= 7200 + 840 + 25 = 8065$ s

Mass

Mass is the measure of how much matter there is in an object. It is often confused with weight (non-scientifically it is often used interchangeably with weight). An object's mass is constant; it would have the same mass on the Earth or on the Moon. It is measured in the SI base unit **kilograms** (kg).

Converting Units of Mass

1 g (gram) = 0.001 kg	To convert from grams to kilograms, multiply by 0.001 (or divide by 1000). To convert from kilograms to grams, divide by 0.001 (which is equivalent to multiplying by 1000).
1 T (tonne) = 1000 kg	To convert from tonnes to kilograms, multiply by 1000. To convert from kilograms to tonnes, divide by 1000.

Example

A cupcake recipe includes 50 g of sugar. Jon doubles the ingredients to make twice as many cupcakes. How many kilograms of sugar will he need?

$50\,g \times 2 = 100\,g$

$100\,g \div 1000 = 0.1\,kg$

Derived Quantities

All other units come from these basic measurements. Dimensional analysis can be very useful for linking equations and units. The key derived units in mechanics are velocity, acceleration and force/weight.

Velocity

Velocity is measured in **metres per second** (ms^{-1}). Velocity is the vector value of speed (i.e. speed but with a set direction). In SI units, it is the number of metres travelled every second. It is the rate of change of displacement with regard to time.

Acceleration

Acceleration is measured in **metres per second squared** (ms^{-2}). Acceleration measures the rate of change of velocity with respect to time.

Force/Weight

Force/weight is measured in **Newtons** (N). Weight is the force an object exerts vertically downwards due to gravity. It is mass multiplied by the acceleration due to gravity. Newtons are equivalent to $kg\,ms^{-2}$, but as force is used so often it is given its own unit of measurement: the Newton (named after Sir Isaac Newton, the 'discoverer of gravity'). Newtons are used to measure all forces, not just those due to gravity. It is the unit force needed to accelerate a mass of 1 kg at a rate of 1 metre per second squared.

Converting Derived Quantities

By considering the separate parts of the derived quantity, it is possible to convert the units.

Example

Convert $14.4\,kmh^{-1}$ into ms^{-1}.

To convert kilometres to metres, multiply by 1000.

To convert hours to seconds, multiply by 3600. As time is in the denominator of the unit ($ms^{-1} = \frac{m}{s}$), this is in effect a division.

$$\frac{km}{h} \xrightarrow{\times 1000 \text{ (top)},\ \times 3600 \text{ (bottom)}} \times\frac{1000}{3600} \rightarrow \frac{m}{s}$$

$14.4\,kmh^{-1} = 14.4 \times \frac{1000}{3600} = 4\,ms^{-1}$

The SI prefixes can be used for these derived measures. The table below shows some of the prefixes that can be used with SI units.

Prefix	'Name'	Multiplier
M	mega	1 000 000
k	kilo	1000
c	centi	0.01
m	milli	0.001
μ	micro	0.000001

Note: Kilograms are the fundamental measure for mass, not grams. Seconds are the fundamental unit for time but since they don't have a relationship in powers of 10 the conversion is slightly more complicated and uses the terms minutes, hours, days, etc., rather than the prefixes above.

Dimensional Analysis

When using equations, they need to make sense in terms of the quantities for each side – this can be a useful way of checking if a formula has been remembered correctly (it doesn't confirm the formula is correct as there may be constants without units, but it can help to spot glaring mistakes). This will not be tested in its own right, but is a useful skill that can support your working.

Example

Rizwan tries to write down the equations of constant acceleration from memory as part of his revision:

$v^2 = u^2 + 2a$ [Equation 1]

$s = ut + \frac{1}{2}at^2$ [Equation 2]

$v = ut + at$ [Equation 3]

By considering the units, show that at least two of these equations have been remembered incorrectly.

Equation 1:

- LHS: v^2 is velocity-squared $\therefore$ measured in $(\text{ms}^{-1})^2 = \text{m}^2\text{s}^{-2}$.
- RHS: first term u^2 is velocity-squared $\therefore$ measured in $(\text{ms}^{-1})^2 = \text{m}^2\text{s}^{-2}$.

  This is consistent with the LHS.
- RHS: second term $2a$ is an acceleration $\therefore$ measured in ms^{-2}.

  This is **not** consistent with the rest of the equation, so it cannot be correct. To make it correct, it would need to be multiplied by a length, as $\text{ms}^{-2} \times \text{m} = \text{m}^2\text{s}^{-2}$. It would then be consistent throughout the terms of the equation.

Equation 2:

- LHS: s is a displacement $\therefore$ measured in m.
- RHS: first term ut is a velocity multiplied by time, $\text{ms}^{-1} \times \text{s} = \text{m}$.

  This is consistent with the LHS.
- RHS: second term $\frac{1}{2}at^2$ is an acceleration multiplied by time-squared, $\text{ms}^{-2} \times \text{s}^2 = \text{m}$.

  This is also consistent with the previous two terms.
- As all the terms have a unit of metres, the equation is balanced dimensionally.

Equation 3:

- LHS: v is a velocity so measured in ms^{-1}.
- RHS: first term ut is a velocity multiplied by time, $\text{ms}^{-1} \times \text{s} = \text{m}$.

  This is **not** consistent with the LHS, so the formula cannot be correct.

The Language of Mechanics

There are many simplifications made in the modelling of real-life situations. It is important to know what the key terminology means in relation to the assumptions made about an object. It is also important to be able to analyse the effect this might have on the theoretical result from the calculation compared to real-life results.

- Body – is used to describe any object with a mass that is being considered.
- Particle – is a modelling assumption meaning that a body has mass but relatively small size. The mass is modelled as acting at a single point. It also means there is no air resistance considered during motion.
  (Note: At AS-level many things are modelled as particles: people, cars, etc.)
- Plane – motion and forces are considered at most in two dimensions. The two-dimensional space is referred to as a plane.
- Light – mass is negligible and so is ignored during calculations.
- Inextensible – the string/rope/cable does not stretch, which means a constant force applied at one end is transferred completely through the string and that the bodies will have the same velocity and acceleration.

- String – a relatively thin cord that joins two bodies. Generally, a string is modelled as light and inextensible.
- Smooth – the surface and the body experience no frictional force.
- Rough – the body and the plane exert a frictional force on each other.

SUMMARY

- **Length is measured in the SI unit metres (m):**

  1000 m = 1 km 0.01 m = 1 cm 0.001 m = 1 mm
- **Time is measured in the SI unit seconds (s):**

  60 s = 1 minute

  3600 s = 60 minutes = 1 hour
- **Mass (not weight !) is measured in kilograms (kg):**

  0.001 kg = 1 g (gram)

  1000 kg = 1 T (tonne)
- **Derived measures include the following:**
  - **Velocity is measured in metres per second (ms^{-1}).**
  - **Acceleration is measured in metres per second squared (ms^{-2}).**
  - **Force/weight is measured in Newtons (N).**
- **You should be comfortable working with derived measures using the conversions above.**
- **Mechanics uses a lot of modelling assumptions. Familiarity with these terms and their meanings is assumed: body; particle; plane; light; inextensible; string; smooth; rough.**

QUICK TEST

1. Convert each of the following into the relevant SI unit:

   a) 2.5 km **b)** 3 hours **c)** 4.2 tonnes
2. Convert each of the following derived quantities into the standard unit shown in the bracket.

   a) A force of 0.34 kN [N]

   b) A velocity of 12 cm per minute [ms^{-1}]

   c) An acceleration of 6480 kmh^{-2} [ms^{-2}]
3. Match each of **A–F** with **P–U**.

A	Light	A connector (string, rope, bar, cord) that doesn't stretch and so transfers the full force applied at one end through to the other end.	P
B	Inextensible	Unchanging, e.g. the variations in the acceleration due to gravity whilst on the Earth are so small that the acceleration due to gravity is considered to be the same anywhere on the Earth.	Q
C	Particle	There is no friction experienced between the body and the plane that the body is travelling along.	R
D	Constant	Modelled as having zero mass, as the mass is negligible compared to other bodies within the set-up.	S
E	Straight line	The body is to be modelled as a point mass.	T
F	Smooth plane	All motion is considered to be running in a single direction (one-dimensional). There needs to be a positive and negative direction defined relative to an origin.	U

Links to Other Concepts

- Forces
- Kinematics
- Integration and differentiation
- Interpreting graphs
- Solving simultaneous equations
- Vectors
- Algebraic manipulation
- Interpreting context-based questions

DAY 5 45 Minutes

Kinematics and Graphs

Kinematics is the study of motion of a body. This can be approached purely algebraically or by considering graphs of motion.

Key Language in Kinematics

- **Position** – where the body is at a set time relative to an origin.
- **Displacement** – how far the body is from the origin at a given time. It is a **vector quantity** so includes a direction as well as a length. The direction is defined by whether the displacement value is positive or negative.
- **Distance** – a scalar quantity. This is the distance between objects where the direction doesn't matter. It can also be used to consider the total **distance travelled** by a body.

Example
If a body travels 2 metres away from the origin, then back to the origin, it would have zero displacement. However, it would have travelled a distance of 4 metres.

- **Velocity** – the rate of change of displacement with respect to time. It is how fast the body is moving at a given time. Velocity is a **vector quantity** and as such has direction, again defined by the positivity/negativity.
- **Speed** – the scalar value of velocity. It is simply how fast the object is travelling but without reference to the direction of travel.
- **Acceleration** – the rate of change of the velocity with respect to time. It is how much the velocity changes each second. Acceleration is also a **vector quantity** so the direction is important and should be consistent with a positive direction for displacement and velocity. **Deceleration** is used to mean slowing down. This should be treated as a negative acceleration in calculations. **Retardation** is a term that means deceleration.

Graphs

An exam question might ask for information to be interpreted and used from a graph, or for a graph to be produced of given information. If asked to sketch a graph, show all key points clearly. If a graph is to be drawn, the plotting should be accurate on the axes provided.

Displacement–Time Graphs

A displacement–time graph shows displacement of the object on the y-axis, with time on the x-axis. The gradient is the rate of change of the displacement with respect to time, which is the **velocity**. As displacement is a vector, displacement can be both positive and negative. To find distance travelled, take the sum of the changes in displacement.

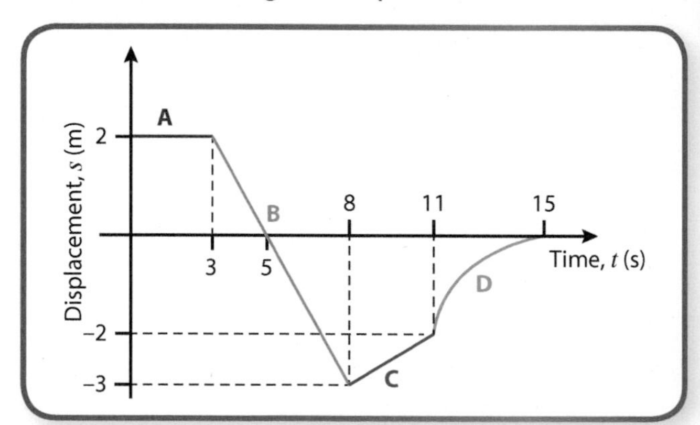

Many displacement–time graphs start from the origin but this isn't always the case, as shown above.

- Section **A** has 0 velocity. The object isn't moving as there is no change in displacement. The gradient of this line segment is 0, so the velocity is 0.
  Note: You will not see vertical line segments on displacement–time graphs as this would represent a body being in multiple positions at the same time.
- Section **B** shows the body moving with constant velocity (as it is a straight line). The velocity is negative as the gradient of the line is negative. The body passes the origin at 5 seconds and continues at the same velocity for a further 3 seconds. To work out the velocity of this section, find the gradient of the line:

$$v = \frac{\text{vertical change}}{\text{horizontal change}} = \frac{-3-2}{8-3} = -1\,\text{ms}^{-1}$$

- Section **C** shows the body moving with a constant positive velocity:

$$v = \frac{\text{vertical change}}{\text{horizontal change}} = \frac{-2-(-3)}{11-8} = \frac{1}{3}\text{ms}^{-1}$$

- Section **D** shows the velocity is changing. It starts quite fast but tends towards 0 as it heads back to the point defined as the origin.

Example

What is the total distance travelled by the body?

Section **A**: distance travelled = 0 m

Section **B**: distance travelled = 5 m (it is displaced −5 m, but since distance is scalar the modulus of +5 is used)

Section **C**: distance travelled = 1 m

Section **D**: distance travelled = 2 m

Total distance travelled = 5 + 1 + 2 = 8 m

When interpreting these types of graph, keep the context in mind.

Example

The graph on page 78 could be given a context. Which of the following contexts is most likely? Justify your answer.

- Context 1: Sarah cycles home from school along a straight road. She passes her house and continues to the shop but realises she has forgotten her purse so she cycles back home.
- Context 2: Sienna is playing golf. She hits the ball along the green past the hole. She hits it back towards the hole then, realising it probably will not go far enough, she uses her club to push it along whilst it is still moving.
- Context 3: Sean is pacing in his office trying to make a decision. He pauses by the window for a few seconds before pacing back across the office. Starting back towards the window, he hears his boss approaching the door and rushes back to his chair, sitting down just as his boss enters.

Analysis of possible contexts:

- Context 1: Sarah would have to live 2 m from school, so is unlikely to choose to cycle. Her house, even if modelled as a point, would be unlikely to fit in the 5 m gap between the school and the shop. The graph would also suggest Sarah cycles at a constant speed. This is highly unlikely but is an assumption used to model journeys in mechanics. Unrealistic times, velocities and distances suggest this isn't a reasonable context for the graph.
- Context 2: The distances are more realistic in this context, however the motion isn't. The velocity of the ball once hit would produce curves, not straight lines, as the ball would slow down from the point when it had been hit. Section D could represent the motion of the ball after being hit. However, the full context is not a reasonable model.
- Context 3: This is the most realistic context. Sean has quite a large office as he can walk 5 m in a straight line, but that isn't unreasonable. If it was 50 m, or 50 cm, then it would be unrealistic. It assumes his pace is constant across the room and his turn is instantaneous – this isn't perfect but is reasonable as a model of the situation.

Velocity–Time Graphs

A velocity–time graph shows the velocity of the object on the y-axis, with time on the x-axis. The gradient of the graph is the rate of change of the velocity with respect to time, which is the acceleration. As velocity is a vector, it can be either positive or negative.

The area under the graph is calculated by multiplying a velocity by time ($\text{ms}^{-1} \times \text{s} = \text{m}$), which gives displacement. If the area is above the x-axis, it is positive displacement; if it is under the x-axis, it is negative displacement. Distance travelled is the total area between the graph and the x-axis taken as a scalar (i.e. all positive) and then added together.

If asked to find areas from a graph without integration, remember the formulae for the areas of a rectangle, a triangle and a trapezium.

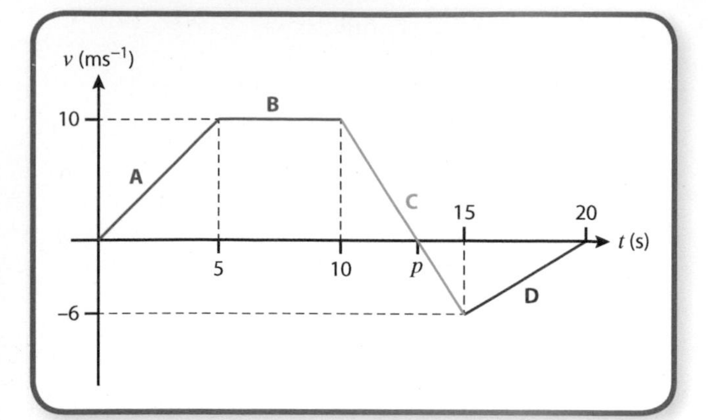

- Section **A** shows the body travelling with a constant, positive acceleration:

$$a = \frac{\text{vertical change}}{\text{horizontal change}} = \frac{10-0}{5-0} = 2\text{ ms}^{-2}$$

- Section **B** has zero acceleration as the velocity is constant. Here the displacement is steadily increasing.
- Section **C** has a constant, negative acceleration (it is decelerating steadily):

$$a = \frac{\text{vertical change}}{\text{horizontal change}} = \frac{-6-10}{15-10} = -\frac{16}{5}\text{ ms}^{-2} = -3.2\text{ ms}^{-2}$$

- Section **D** has a constant, positive acceleration:

$$a = \frac{\text{vertical change}}{\text{horizontal change}} = \frac{0-(-6)}{20-15} = \frac{6}{5}\text{ ms}^{-2} = 1.2\text{ ms}^{-2}$$

Problem-solving skills may be needed to find a missing value.

Example

a) Find the value of p and hence the distance travelled by the object in time p seconds.

There are two possible options for finding p. One is to use the gradient for section C and use properties of straight lines to solve when $y = 0$. The other is to use similar triangles.

Equation of line section C is $y = mx + c$

$10 = -3.2 \times 10 + c$

$c = 10 + 32 = 42$

$y = -3.2x + 42$

p is the value of x when $y = 0$: $0 = -3.2x + 42$

$3.2x = 42$

$x = 13.125$

To find the distance travelled in this time, use the area under the graph:

Area of trapezium $= \frac{1}{2}(a+b)h$

$= \frac{1}{2}(5+13.125)\times 10$

$= 90.625\text{m}$

b) Find the displacement of the object from the origin after 20 seconds.

Displacement to time $p = 90.625$ m

Displacement between p and 20 = area of the triangle

$= \frac{1}{2}bh$

$= \frac{1}{2}\times(20-13.125)(-6)$

$= -20.625\text{m}$

Displacement $= 90.625 + -20.625 = 70\text{m}$

c) Find the distance travelled by the object in the first 20 seconds.

Distance travelled to time p = 90.625 m

Distance travelled between p and 20 = area of the triangle

$= \frac{1}{2}bh$

$= \frac{1}{2} \times (20-13.125) \times 6$

Note: As distance is scalar the modulus is taken, i.e. the positive.

$= 20.625$ m

Total distance travelled $= 90.625 + 20.625$

$= 111.25$ m

Sketching a Displacement–Time Graph

- If given a velocity, it is the gradient of the graph. Stationary means zero velocity so the line segment would be horizontal.
- Use given points (e.g. starting at the origin, at time 5 seconds it is 4 m from the origin). Remember that the graph will always join together and never be a vertical line.
- Questions with a context tend to get very wordy and can be rather complicated to follow. Re-read, make notes and highlight relevant information.

- If representing two bodies' motion on a displacement–time graph, they are at the same place at the same time where the lines intersect.

Example

A car is travelling due East along a straight, horizontal road with a velocity of $14\,\text{ms}^{-1}$. At time $t = 0$ the car passes point P. After 3 seconds the car stops at a set of traffic lights where it has to wait for 20 seconds. The car then continues with a constant velocity and reaches its destination, point Q, which is 340 m East of P at $t = 43$ seconds.

a) Sketch a displacement–time graph for its motion.

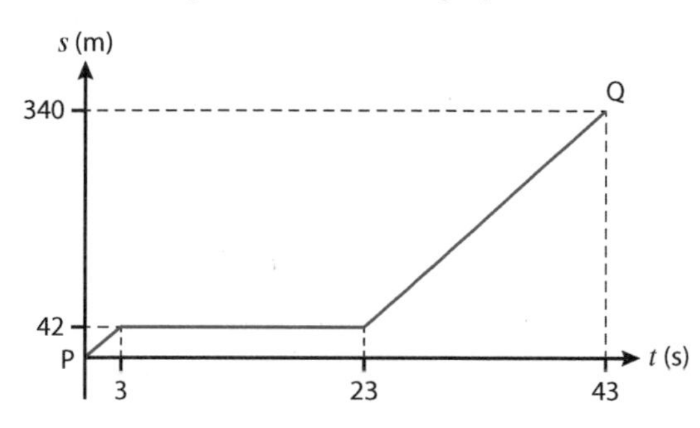

b) Find the velocity of the car after the traffic lights.

Gradient = velocity

$$= \frac{340 - 42}{20} = 14.9\,\text{ms}^{-1}$$

c) The road has a speed limit of 40 miles per hour. Does the car stay within the speed limit? Use 1609.34 metres = 1 mile.

$$\begin{aligned}40 \text{ miles per hour} &= 40 \times 1609.34 \text{ metres per hour}\\ &= 64\,373.6 \div 3600\\ &= 17.881\ldots\,\text{ms}^{-1}\end{aligned}$$

The car stays within the speed limit.

Sketching a Velocity–Time Graph

- The acceleration is the gradient of the graph.
- If given a displacement or distance travelled, it is the area under the graph.
- If representing two bodies on a velocity–time graph, the intersection between the two lines shows where the bodies have the same velocity at the same time.
- To find where the two bodies meet means equating the areas underneath them – they may well be travelling at different speeds but will have the same displacement at the same time.

Example

A body is moving along a straight line with a constant velocity of $v = -4\,\text{ms}^{-1}$ for a seconds. It then accelerates with a constant acceleration of $2\,\text{ms}^{-2}$ until it is back at its starting point, at time t.

a) Sketch a velocity–time graph of its motion.

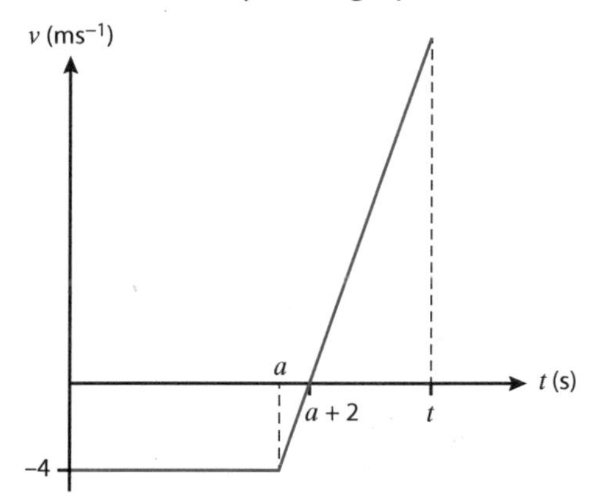

A constant velocity is a horizontal line. When given unknowns instead of numbers, they should still be labelled on the graph. Don't rely on a sketch for estimating times. As the graph crosses the x-axis, the body is momentarily stationary, then it starts to accelerate back towards the origin.

b) Find an equation for t in terms of a in the form $t = b\sqrt{a+1} + (a+2)$.

The distance travelled negatively away from the origin is equal to the distance travelled positively back towards the origin. Find the areas of the shapes in terms of a and t:

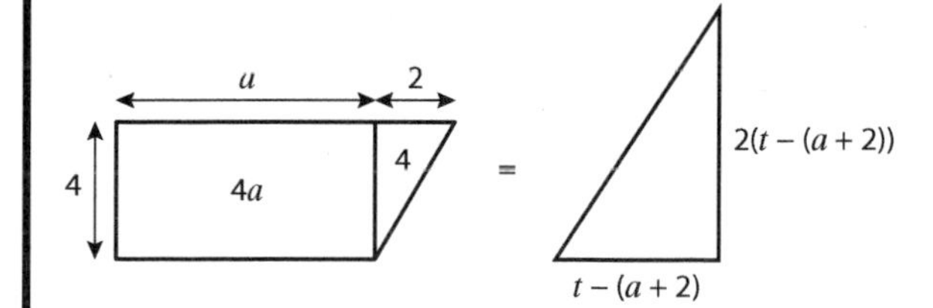

$$4a + 4 = (t - (a+2))^2$$

$$2\sqrt{a+1} = t - (a+2)$$

$$t = 2\sqrt{a+1} + (a+2)$$

c) Given that $a = 15$, find the maximum velocity of the body.

Maximum velocity is when time $= t$

$t = 15 + 2 + 2\sqrt{15+1}$

$= 17 + 2\sqrt{16} = 25\text{s}$

$v = 2(t - (a+2)) = 2(25 - (15+2))$

$= 16\text{ms}^{-1}$

d) A second body starts from rest at the origin and accelerates constantly at 0.5 ms^{-2}. Both bodies continue with their constant acceleration beyond 25 seconds. Find the time when the two bodies are travelling with the same velocity and state the velocity.

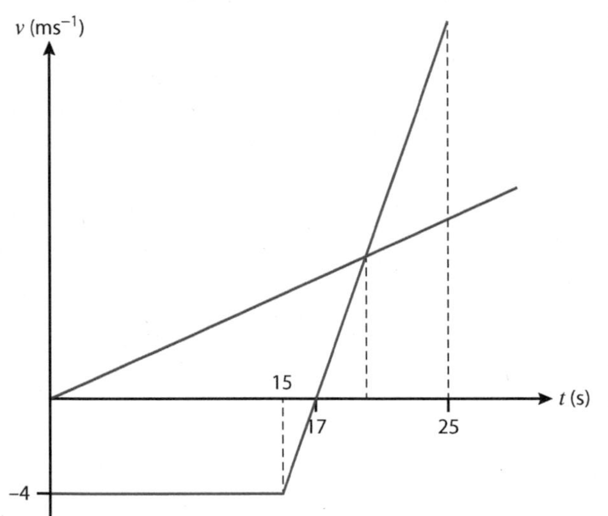

This is the point where the two lines intersect.

The second body is a straight line through the origin with gradient 0.5, so $v = 0.5t$

The first body will be on the diagonal line, which has gradient 2 and goes through point (17, 0), so $v = 2t - 34$

Equating the two lines: $0.5t = 2t - 34$

$1.5t = 34$

$t = \frac{34}{1.5} = \frac{68}{3} = 22\frac{2}{3}\text{ s}$

At $22\frac{2}{3}$ s, both bodies are moving at $11\frac{1}{3}\text{ms}^{-1}$.

e) At what time do the two bodies pass each other?

They pass each other when the displacement from the origin is equal.

Displacement of first body =

$-64 + \frac{1}{2} \times (t-17) \times 2(t-17) = t^2 - 34t + 225$

Displacement of second body $= \frac{1}{2}t(0.5t) = 0.25t^2$

Equal displacement when $0.25t^2 = t^2 - 34t + 225$

$3t^2 - 136t + 900 = 0$

$t = \frac{136 \pm \sqrt{(-136)^2 - 4 \times 3 \times 900}}{2 \times 3}$

$t = 37.28\ldots$ or $t = 8.04\ldots$

As $t > 17$, $t = 37.3$ s (3 s.f.)

Links to Other Concepts

- Coordinate geometry
- Solving quadratics
- Integration and differentiation
- Solving simultaneous equations
- Vectors
- Algebraic manipulation

SUMMARY

- **Displacement–time graphs:**
  - ***y*-value is displacement at time *t*.**
  - **Gradient is the rate of change of displacement with respect to time, i.e. velocity.**
  - **Horizontal line means the object is stationary.**
  - **If two lines intersect, the two objects are in the same place at the same time.**
  - **Curves represent varying velocity.**
- **Velocity–time graphs:**
  - ***y*-value is velocity at time *t*.**
  - **Gradient is the rate of change of velocity with respect to time, i.e. acceleration.**
  - **Horizontal line means the object is travelling at a constant velocity.**
  - **Area between the graph and the *x*-axis is the displacement; beneath the *x*-axis it is negative displacement.**
  - **Total area (ignoring negatives) gives the total distance travelled.**
  - **If two graphs intersect, the two objects have the same velocity at that time.**
  - **To find when two objects are at the same place, equate the displacements (area under graph) and be careful to include the negatives if relevant.**
- **Sketches can help to model a situation but don't rely on them for exact answers.**

QUICK TEST

1. What is the name of speed with a given direction?
2. 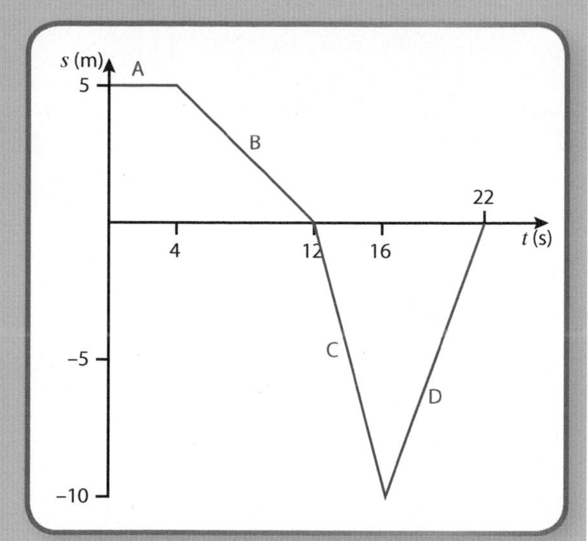

   a) What is the velocity during section D?
   b) What is the highest speed experienced during the motion shown?
   c) At what times is the object at the origin?
   d) Describe what is happening in section A.
3. 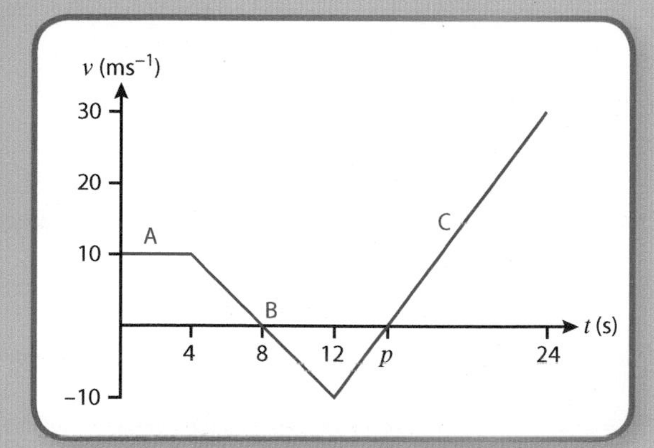

   a) What is the acceleration during section A?
   b) What is the acceleration during section B?
   c) What is the acceleration during section C?
   d) What is the displacement of the object during the first 4 s?
   e) What is the value of p?
   f) What is the total distance travelled in 24 s?
   g) What is the final displacement at 24 s?

PRACTICE QUESTIONS

1. The velocity–time graph shows a motorcyclist (Biker 1) travelling along a straight, horizontal road.

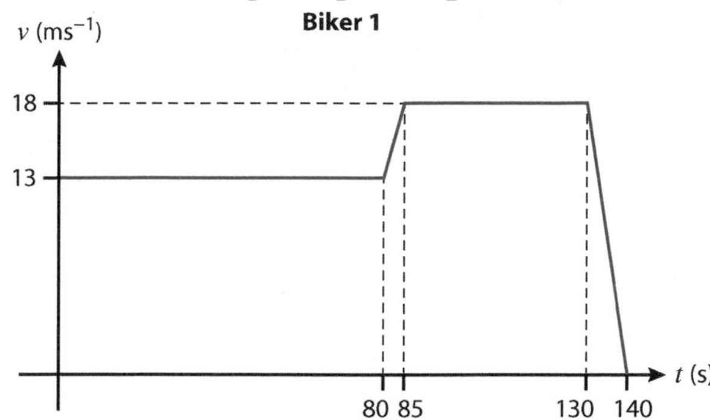

At $t = 0$, Biker 1 passes point A on the road. At point B, Biker 1 passes a change of speed limit sign and accelerates to the new speed limit.

   a) What is the new speed limit, in miles per hour? Use 1610 metres = 1 mile. **[2 marks]**
   b) How far is the speed limit sign from A? **[1 mark]**

A second motorcyclist (Biker 2) leaves A at $t = 0$. The velocity–time graph shows his journey.

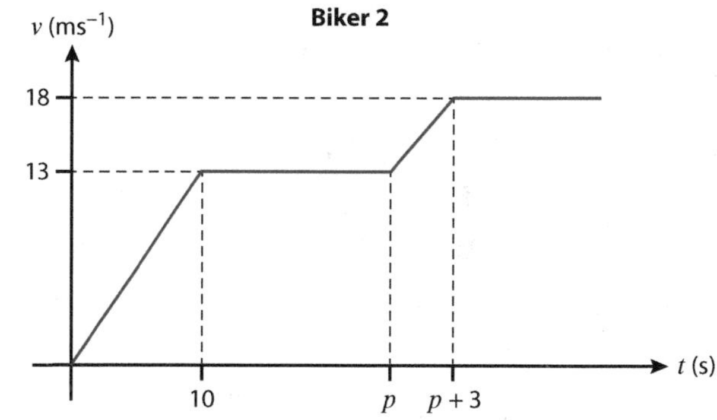

   c) Given that Biker 2 also adheres to the speed limit, find the time when they pass point B. **[4 marks]**

2. A train travels along a straight, horizontal track from point P to point Q. For the first 202.5 m it accelerates, from rest, with a constant acceleration of 5 ms^{-2}. It then travels with a constant velocity for a further 2.07 km. As the train approaches a built-up area, it decelerates to 20 ms^{-1}, taking 5 seconds to do so. The train continues at 20 ms^{-1} until it passes point Q, 1.5 minutes after leaving P.
   a) Find the distance PQ. **[7 marks]**
   b) The train starts to decelerate constantly at Q and comes to a complete stop a total distance of 3.1 km from P. Find the deceleration of the train in this last period of motion. **[2 marks]**

Vectors 1

A vector is a measure which has both **magnitude** (size) and **direction**. If navigating on a ship, being told 'the port is 2 km away' doesn't help get to the port without further information. The distance is the magnitude, the 'size' or 'amount' of the measure; these are **scalars**. To be useful, you would also need to know the direction. 'The port is 2 km due West' is much more useful. By adding a direction to the scalar quantity, it becomes a **vector**.

Vectors are often used to describe positions of objects. At AS-level this is on a 2D plane. Other key vector quantities encountered at AS-level are force, velocity and acceleration (also only applied in 2D).

Describing Vectors

A vector can be described in a number of ways, as long as the description gives both size and direction.

Example
'Due East for 5 km.'

'A force of 5 N at an angle of 30° to the horizontal.' This could also be written using magnitude-direction form (r, θ) as: (5 N, 30°).

A vector can be drawn as an arrow, with the length representing the magnitude of the vector and the arrow itself, along with the orientation of the line, defining the direction.

A vector that represents movement between two points is called a displacement vector.

The vector representing a movement from O to A can be written as $\overrightarrow{OA}$.

It is also possible to define a name for a vector. Generally lowercase letters (a, b, c, d, … or p, q, r, s, …) are used. When handwritten, they are underlined, e.g. $\underline{a}$, to show that it represents a vector and not just a number. In print, vector values are represented in bold.

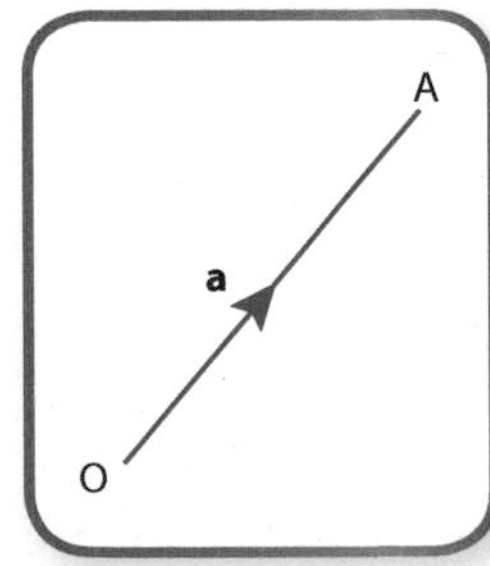

Vector Components

Much as we describe points on a Cartesian graph, we can describe a vector by splitting it into two components which act at right angles to each other. This is known as resolving a vector into its components. As they are perpendicular, they do not affect each other. Generally this is based on a plane oriented with one component of the direction being horizontal or to the right (negative meaning to the left) and the other being vertical or upwards (negative being downwards); or East (negative West) and North (negative South).

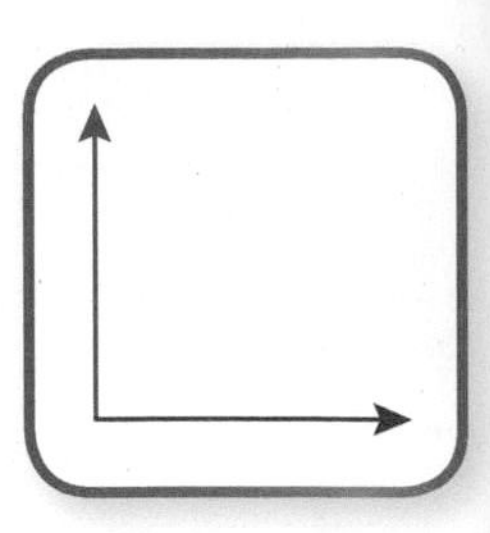

Writing Vectors in Component Form: i and j Notation

The **i** and **j** represent a unit vector in a given direction. Usually they would be considered as follows, though it is possible to define them differently. Importantly, the two directions must be perpendicular:

i	One space to the 'right'	One space East
j	One space 'up'	One space North

The vector can be multiplied by a constant value, which changes its magnitude but not its direction (unless the coefficient is negative, in which case the direction is reversed).

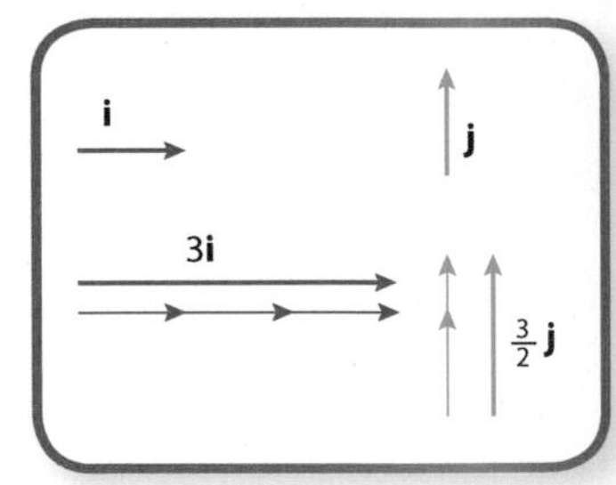

Example
Describe the vectors **a**, **b** and **c**, as shown on the grid, using **i** and **j** vector notation.

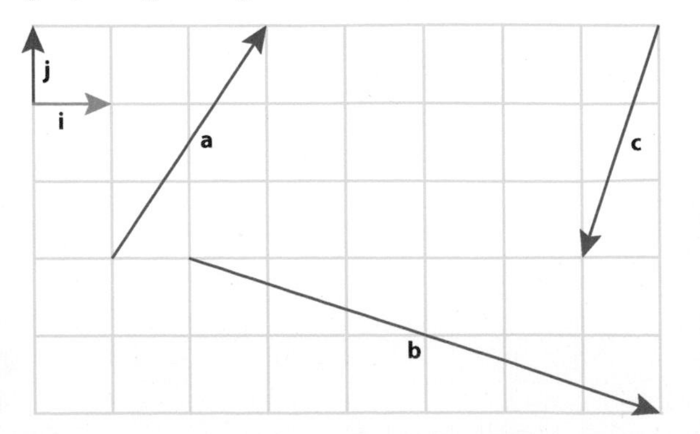

a: The line goes two spaces right (+2**i**) and three spaces up (+3**j**), so **a** = 2**i** + 3**j**

b: The line goes six spaces right (+6**i**) and two spaces down (–2**j**), so **b** = 6**i** – 2**j**

c: The line goes one space left (–**i**) and three spaces down (–3**j**), so **c** = –**i** – 3**j**

Column Vector Notation

Column vector notation is used to describe vertical and horizontal translations on a graph. It is a vertical bracket containing two numbers: $\begin{pmatrix} a \\ b \end{pmatrix}$, where a represents the 'horizontal' change (positive x-direction on a graph) and b the 'vertical' change (positive y-direction on a graph).

Both column vector notation and **i** and **j** notation represent the same idea and they can be used interchangeably, unless the question specifies the notation required.

Example
Describe the vectors **a**, **b** and **c**, as shown on the grid, using column vector notation.

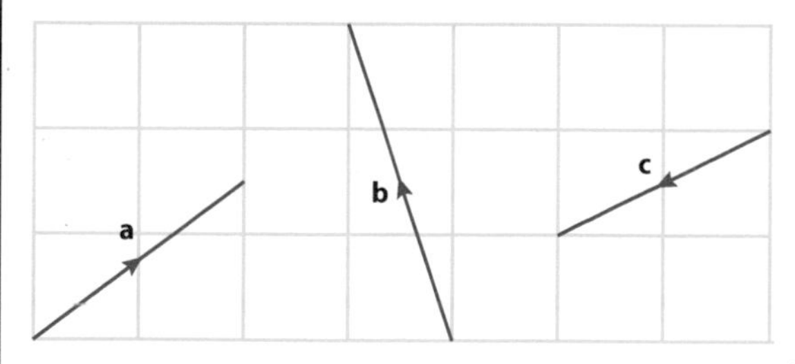

a: The line goes two spaces right and one-and-a-half spaces up: $\mathbf{a} = \begin{pmatrix} 2 \\ \frac{3}{2} \end{pmatrix}$

b: The line goes one space left and three spaces up: $\mathbf{b} = \begin{pmatrix} -1 \\ 3 \end{pmatrix}$

c: The line goes two spaces left and one space down: $\mathbf{c} = \begin{pmatrix} -2 \\ -1 \end{pmatrix}$

Finding the Vector Between Coordinates

To find the displacement vector, $\overrightarrow{AB}$, between the points A (a, b) and B (c, d):

- Horizontal change goes from a to c and is found using $c - a$
- Vertical change goes from b to d and is found using $d - b$.

Example
Find the displacement vector between the points A and B with coordinates (2, 3) and (1, 7) respectively.

Horizontal change = 1 – 2 = –1

Vertical change = 7 – 3 = 4

$\overrightarrow{AB} = \begin{pmatrix} -1 \\ 4 \end{pmatrix}$

Finding Magnitudes from Components

To find the magnitude of a vector, use Pythagoras' Theorem, $a^2 = b^2 + c^2$, where a is the magnitude of the vector and b and c are the **i** and **j** components. Two vertical lines on either side of a vector denote 'the magnitude'.

Example
a = 4**i** + 3**j**

$|\mathbf{a}| = \sqrt{4^2 + 3^2}$

$= \sqrt{16+9} = \sqrt{25} = 5$

As the value found is scalar, representing size only, the answer is positive.

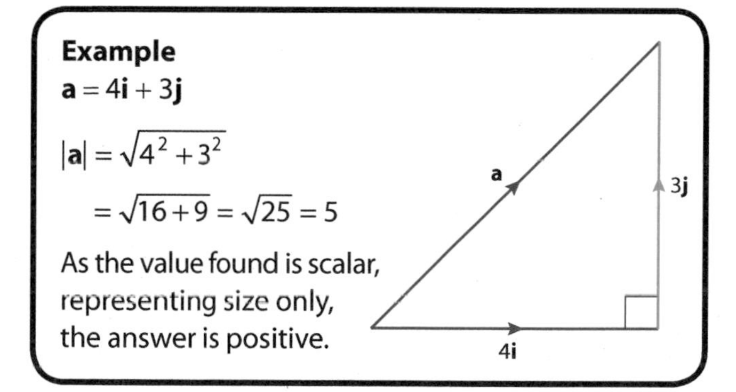

Defining Direction from Components

As shown above, the components form a right-angled triangle with the vector. To define direction an angle is needed, so simple trigonometric ratios are used. Be careful to find the correct angle, as defined by the question.

Example

The vector $\mathbf{b} = \begin{pmatrix} 2 \\ -5 \end{pmatrix}$. Find the angle between the vector and the unit vector $\begin{pmatrix} 0 \\ 1 \end{pmatrix}$ (often called **j**).

$\tan\alpha = \frac{\text{opposite}}{\text{adjacent}} = \frac{5}{2}$

$\alpha = \tan^{-1}\left(\frac{5}{2}\right)$

$= 68.1985\ldots$

$\theta = \alpha + 90 = 158.2°$ (1 d.p.)

Resolving Vectors into Components

If given the magnitude and the direction of a vector, you can resolve it into components and express it in either of the forms outlined above.

Example

A ship sails on a bearing of 030° for a distance of 6 km. Express the vector as a column vector.

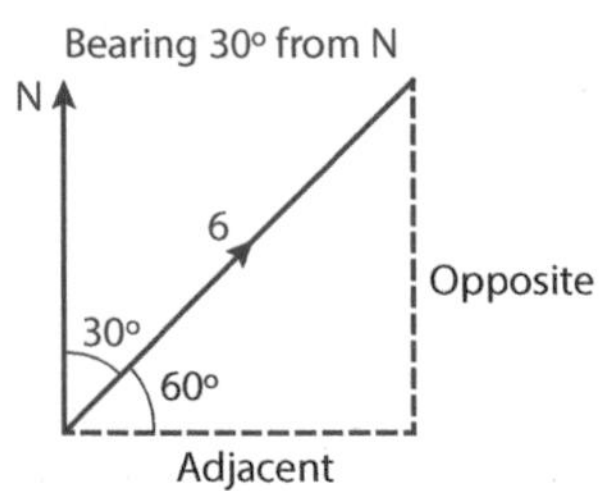

Horizontal (East) component $= 6\cos 60$

$= 6 \times \frac{1}{2}$

$= 3$

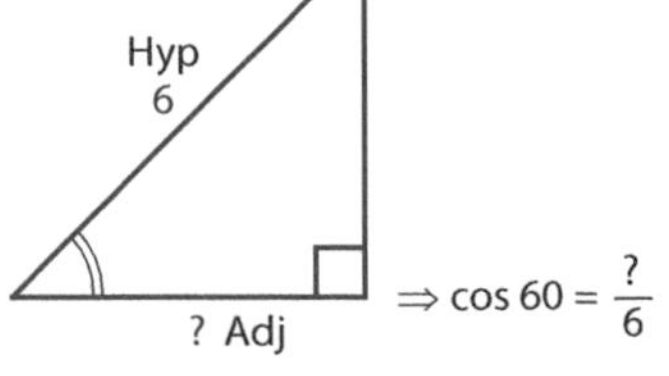

Vertical (North) component $= 6\sin 60$

$= 6 \times \frac{\sqrt{3}}{2}$

$= 3\sqrt{3}$

Column vector $\begin{pmatrix} 3 \\ 3\sqrt{3} \end{pmatrix}$

SUMMARY

- **Vectors have both magnitude and direction. Scalar values have only magnitude.**
- **Vectors are represented by a line with an arrow on, denoting the direction.**
- **i and j are unit vectors in the horizontal and vertical directions.**
- **Component form of a vector:**

  p = ai + bj is a vector that represents a units to the right and b units upwards.

  p can also be represented by the column vector $\begin{pmatrix} a \\ b \end{pmatrix}$.
- **To find the magnitude of a vector, |p|, use Pythagoras' Theroem. $|\mathbf{p}| = \sqrt{a^2 + b^2}$**

  Look out for magnitude being hidden as 'speed', 'distance' or another associated scalar value.
- **To find the angle:**
  - **Sketch a diagram and take care identifying which angle is being asked for and how to find it.**
  - **Use trigonometry (SOH CAH TOA) to calculate the angle.**
- **To find the components:**
  - **Sketch a diagram, carefully labelling the given angle.**
  - **Use trigonometry (SOH CAH TOA) to find the component vectors.**
- **Use the form specified by the question. If no form is specified, use the one that is easiest for you.**

Links to Other Concepts

- Forces and Newton's laws
- Motion in a straight line
- Velocity/speed, displacement/distance
- Trigonometry
- Geometry
- Coordinates
- Problem solving
- Simultaneous equations

QUICK TEST

1. State the value of tan(45).
2. Find θ where $\tan\theta = \frac{4}{3}$
3. The vector $\mathbf{p} = \begin{pmatrix} -2 \\ \frac{3}{4} \end{pmatrix}$
   a) Convert **p** into **i** and **j** notation.
   b) Sketch the vector **p**.
   c) Find the magnitude of the vector **p**.
   d) What angle does **p** make with the unit vector **i**?
4. Write each of the following vectors using a column vector and using **i** and **j** notation.

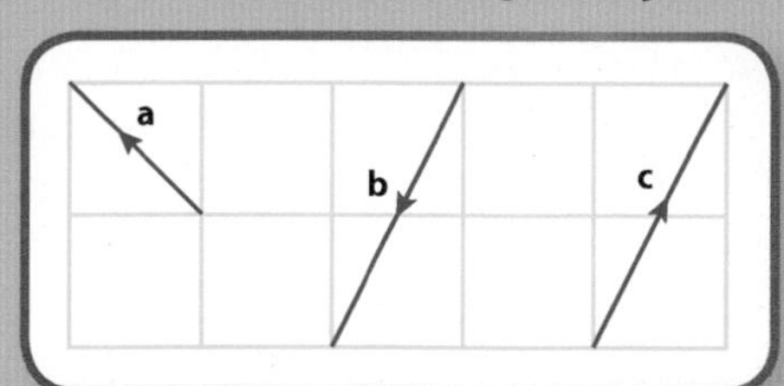

5. The displacement vector **p** represents the movement from point A (–3, 12) to point B (5, 3). Write **p** as a column vector.
6. A force of $3\sqrt{2}$ N acts at an angle of 45° to the horizontal. Express the vector using **i** and **j** notation, where **i** is a unit vector in the horizontal direction.

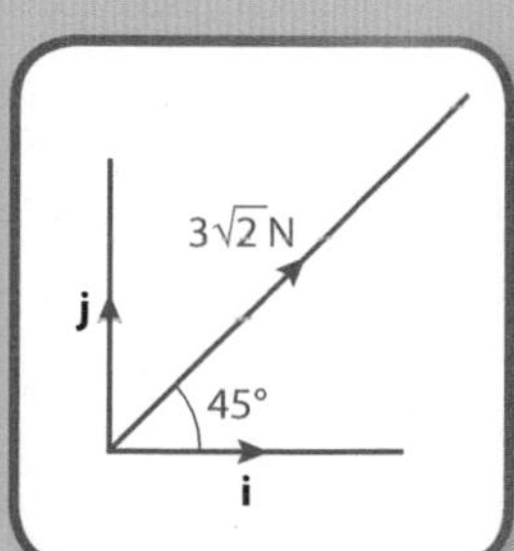

7. An aeroplane flies on a bearing of 240° at a speed of 850 kmh^{-1}. Express the plane's velocity as a column vector.

PRACTICE QUESTIONS

1. The points A and B have coordinates (7, –2) and (2, 10) respectively.
   a) Write down $\overrightarrow{AB}$ as a column vector. **[2 marks]**
   b) Find $|\overrightarrow{AB}|$. **[2 marks]**
   c) What is the angle between $\overrightarrow{AB}$ and the positive x-axis? **[2 marks]**
2. A ball bearing, which weighs 0.01 kg, is sitting on a smooth, horizontal plane. A force of 0.5 N is applied at an angle of 30° to the positive x-direction. Given that acceleration $a = p\mathbf{i} + q\mathbf{j}$, find the values of p and q.
   $a = \frac{F}{m}$ **[4 marks]**
3. A ship travels along a vector $\overrightarrow{AB} = 2a\mathbf{i} + (11 - a)\,\mathbf{j}$ km, where a is a postive number, followed by a vector $\overrightarrow{BC} = 6\mathbf{i} - 8\mathbf{j}$ km. It travels a total distance of 23 km.
   a) Find the overall displacement vector $\overrightarrow{AC}$. **[6 marks]**
   b) What is the exact distance between points A and C? **[2 marks]**

Vectors 2

Adding Vectors

The resultant vector (i.e. the overall vector when the individual vectors are added together) can be found by placing the individual vectors tail to point. By drawing a diagram, you can see where the right-angled triangles are and use them to find the values of magnitude and direction for the resultant vector.

Example

Draw the resultant vector of **a** and **b**.

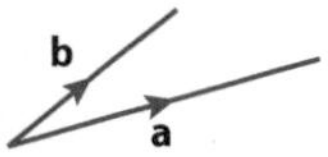

Resultant vector:

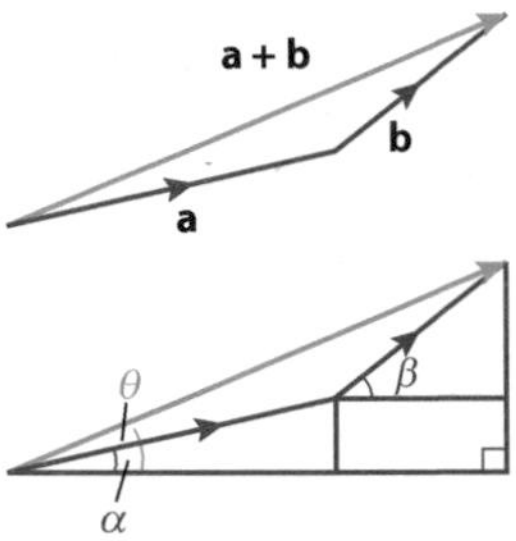

A common context is force addition and finding the resultant force on a body.

Example

A body, modelled as a particle on a smooth, horizontal plane, experiences a force of 3 N at an angle of 30° to the horizontal, and a force of 4 N downwards, as shown in the diagram.

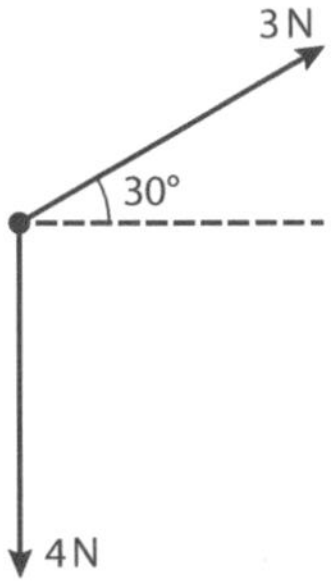

What is the resultant force acting on the body?

Redrawing to represent the vector addition gives:

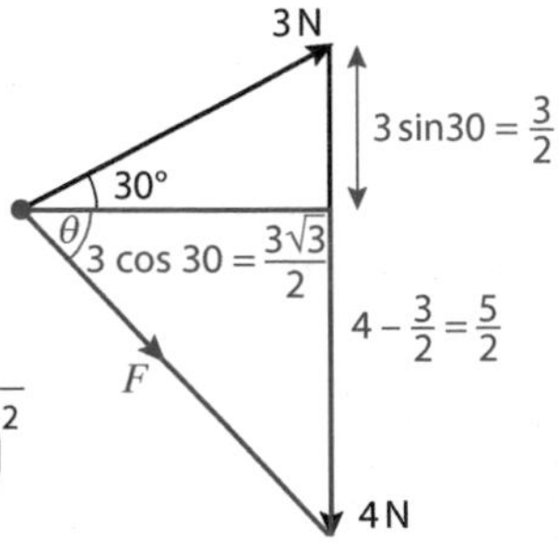

This shows that the force F has

$$\text{magnitude} = \sqrt{\left(\frac{3\sqrt{3}}{2}\right)^2 + \left(\frac{5}{2}\right)^2}$$

$$= \sqrt{13}$$

At an angle of $\tan^{-1}\left(\frac{5}{3\sqrt{3}}\right) = 43.89788 \ldots$

$= 43.9°$ (1 d. p.) below the horizontal.

Multiplication by Scalars

A vector can be multiplied by constants to create a vector that is effectively enlarged (which can mean getting smaller) by a scale factor. Adding two lots of vector **a** could be written as 2**a**. If the vector is defined by its components (either **i** and **j**, or column vector), each component is multiplied by the constant.

Example

$$\mathbf{a} = \begin{pmatrix} 5 \\ -2 \end{pmatrix}$$

$$2\mathbf{a} = 2 \times \begin{pmatrix} 5 \\ -2 \end{pmatrix} = \begin{pmatrix} 2 \times 5 \\ 2 \times -2 \end{pmatrix} = \begin{pmatrix} 10 \\ -4 \end{pmatrix}$$

Adding Vectors – Algebraically

A defined vector can be added to, and subtracted from, other vectors. You can imagine vectors as small sections of pathway that can be laid end to end to make a different pathway. The important thing is the start and end points, rather than the specific route. Subtraction would mean reversing the direction suggested by the arrow.

Example

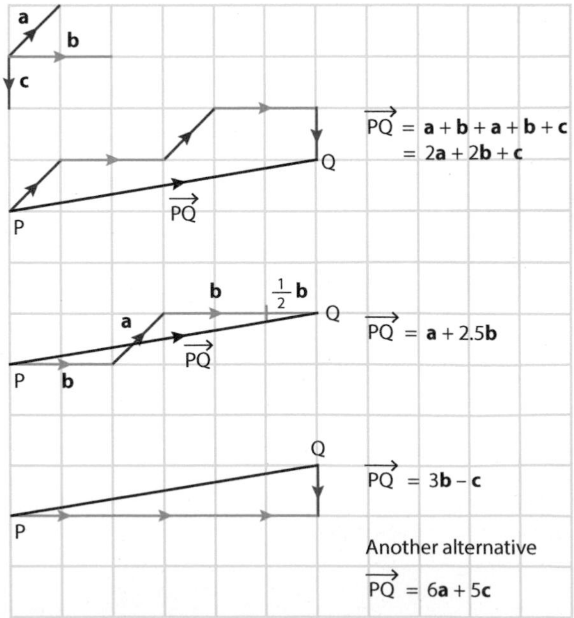

With the vectors defined in the previous example, there are multiple ways of getting from P to Q. This is because the vectors can be used to represent each other, i.e. $\mathbf{b} = 2\mathbf{a} + 2\mathbf{c}$. If you had only vectors **a** and **b**, there would still be multiple pathways that could be created by using different orders, but they would all simplify to the same $\mathbf{a} + 2.5\mathbf{b}$, and the overall vector would be equal.

Geometrical Problem Solving

Problems will often require a certain amount of interpretation and knowledge of geometrical shapes and their properties. Path finding is important here.

Example

Regular hexagon ABCDEF has a vector $\mathbf{u} = \overrightarrow{AB}$ and $\mathbf{v} = \overrightarrow{AD}$.

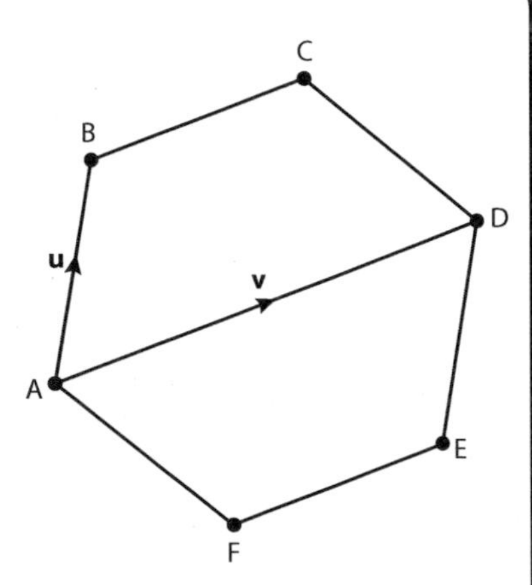

Using these vectors, state the following:

a) $\overrightarrow{ED}$

E to D is the same length and direction as A to B, so $\overrightarrow{ED} = \mathbf{u}$

b) $\overrightarrow{FE}$

F to E is half the length of A to D but is in the same direction, so $\overrightarrow{FE} = \frac{1}{2}\mathbf{v}$

c) $\overrightarrow{AF}$

To go from A to F, consider going to the midpoint of the hexagon then down to F. That is, along $\frac{1}{2}\overrightarrow{AD}$ then the equivalent of $\overrightarrow{BA} = -\overrightarrow{AB}$

$\overrightarrow{AF} = -\mathbf{u} + \frac{1}{2}\mathbf{v}$

Parallel vectors are in the same direction. To prove that vectors are multiples of each other is to prove that they are parallel. If told they are parallel, it implies that they are multiples of each other.

Example

$\overrightarrow{AB} = 5\mathbf{i} - 3\mathbf{j}$ and $\overrightarrow{CD} = 2.5\mathbf{i} - 1.5\mathbf{j}$. Show that the vectors $\overrightarrow{AB}$ and $\overrightarrow{CD}$ are parallel.

$\overrightarrow{AB} = 5\mathbf{i} - 3\mathbf{j} = 2(2.5\mathbf{i} - 1.5\mathbf{j})$

$\overrightarrow{AB} = 2\overrightarrow{CD}$ so the vectors are parallel.

As a 'show that' question is a proof-based one, all steps need to be shown clearly and in detail.

Example

The vectors $\begin{pmatrix} 5 \\ -2 \end{pmatrix}$ and $\begin{pmatrix} p \\ -6 \end{pmatrix}$ are parallel. Find the value of p.

For the vectors to be parallel $\begin{pmatrix} p \\ -6 \end{pmatrix} = k\begin{pmatrix} 5 \\ -2 \end{pmatrix}$

$\begin{pmatrix} p \\ -6 \end{pmatrix} = \begin{pmatrix} 5k \\ -2k \end{pmatrix}$

$-6 = -2k \quad \rightarrow \quad k = 3$

$p = 5k = 5 \times 3 \quad \rightarrow \quad p = 15$

Collinearity is when three or more points lie on the same straight line. To prove collinearity of points A, B and C, the vectors $\overrightarrow{AB}$ and $\overrightarrow{AC}$ need to be shown to be parallel (i.e. be multiples of each other). If they have a point in common and have the same direction, the three points must lie on a straight line.

Example

$\overrightarrow{AB} = -1\mathbf{i} - 3\mathbf{j}$ $\quad$ $\overrightarrow{BC} = -2\mathbf{i} - 6\mathbf{j}$

Show that A, B and C are collinear.

$\overrightarrow{AC} = \overrightarrow{AB} + \overrightarrow{BC} = -1\mathbf{i} - 3\mathbf{j} + -2\mathbf{i} - 6\mathbf{j}$

$= -3\mathbf{i} - 9\mathbf{j}$

$= 3(-1\mathbf{i} - 3\mathbf{j})$

$\overrightarrow{AC} = 3 \times \overrightarrow{AB}$, so A, B and C are collinear.

Example

The vector diagram shows points A, B, C, D and O.

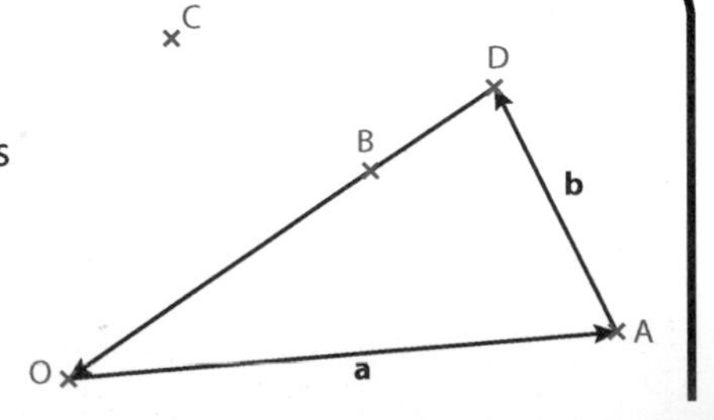

$\overrightarrow{OA} = \mathbf{a}$, $\overrightarrow{AD} = \mathbf{b}$ and $\overrightarrow{OC} = \frac{1}{2}\mathbf{a} + \frac{3}{2}\mathbf{b}$

Given that A, B and C are collinear, find the ratio OB : OD.

$\overrightarrow{AC} = -\mathbf{a} + \overrightarrow{OC} \quad = -\frac{1}{2}\mathbf{a} + \frac{3}{2}\mathbf{b}$

As A, B and C are collinear:

$\overrightarrow{AB} = k\left(-\frac{1}{2}\mathbf{a} + \frac{3}{2}\mathbf{b}\right) = -\frac{k}{2}\mathbf{a} + \frac{3k}{2}\mathbf{b}$

Let p be fraction $\frac{\overline{OB}}{\overline{OD}}$

$\overrightarrow{OB} = p\overrightarrow{OD}$

$\overrightarrow{AB} = \overrightarrow{AO} + p\overrightarrow{OD}$

$= -\mathbf{a} + p(\mathbf{a} + \mathbf{b}) = (p - 1)\mathbf{a} + p\mathbf{b}$

Equate coefficients:

a: $p - 1 = -\frac{k}{2}$ **b**: $p = \frac{3k}{2}$

$p = 1 - \frac{k}{2}$

Substitute in variable: $1 - \frac{k}{2} = \frac{3k}{2}$

$1 = \frac{4k}{2} \rightarrow 2k = 1 \rightarrow k = \frac{1}{2}$

$p = 1 - \frac{\frac{1}{2}}{2} \rightarrow p = \frac{3}{4}$

OB : OD = 3 : 4

Position Vectors

Position vectors relate the position of a point to an origin, much like giving a pair of coordinates on a set of axes.

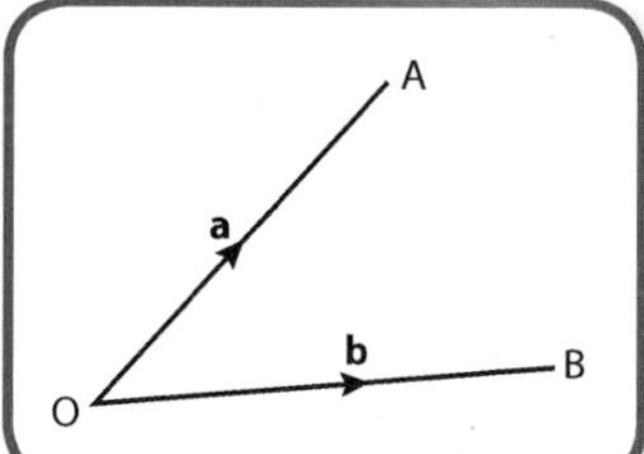

You can see that to go from A to B along predefined routes would be $-\mathbf{a} + \mathbf{b}$.

To find the position vector of B from A:
$\overrightarrow{AB} = \overrightarrow{OB} - \overrightarrow{OA}$

Example

The points A and B have position vectors $\overrightarrow{OA} = -2\mathbf{i} - 6\mathbf{j}$ and $\overrightarrow{OB} = p\mathbf{i}$.

a) Find, in terms of p, the vector $\overrightarrow{BA}$.

$\overrightarrow{BA} = \overrightarrow{OA} - \overrightarrow{OB} = -2\mathbf{i} - 6\mathbf{j} - p\mathbf{i}$

$= (-2 - p)\mathbf{i} - 6\mathbf{j}$

b) Given that the distance BA = 6.5 km, find the possible values of p.

Using Pythagoras' Theorem (other methods are possible):

$\left(\left|\overrightarrow{BA}\right|\right)^2 = (-2 - p)^2 + (-6)^2$

$6.5^2 = 4 + 4p + p^2 + 36$

$\frac{169}{4} = p^2 + 4p + 40$

$4p^2 + 16p + 160 = 169$

$4p^2 + 16p - 9 = 0$

$(2p - 1)(2p + 9) = 0$

$p = \frac{1}{2}$ or $p = -\frac{9}{2}$

SUMMARY

- **Vectors can be added, subtracted and multiplied by constants much like an algebraic term.**
  - **If the vectors are in component form, you can add the components separately, i.e.** $\begin{pmatrix} a \\ b \end{pmatrix} + \begin{pmatrix} c \\ d \end{pmatrix} = \begin{pmatrix} a + c \\ b + d \end{pmatrix}$
- **If multiplying by a constant, multiply each component by the constant. This will change the magnitude of the vector but not the orientation.**

  $k \times \begin{pmatrix} a \\ b \end{pmatrix} = \begin{pmatrix} ka \\ kb \end{pmatrix}$

  Note: A negative scale factor would reverse the vector's direction.
- **Position vectors relate points in space, often to a fixed origin O. To find a relative position vector, use:** $\overrightarrow{AB} = \overrightarrow{OB} - \overrightarrow{OA}$
- **For vectors to be parallel, they must be a multiple of each other.**
- **For points to be collinear, the vectors that join one point to each of the others must be multiples of each other (this can be a negative multiple).**

Links to Other Concepts

- Forces
- Kinematics
- Trigonometry
- Pythagoras' Theorem
- Geometry
- Ratios and fractions
- Algebra, simplification
- Graph transformations

QUICK TEST

1. Find the resultant vector of these pairs:

   a) $\mathbf{a} = \begin{pmatrix} 3 \\ 2 \end{pmatrix}$ and $\mathbf{b} = \begin{pmatrix} -3 \\ 5 \end{pmatrix}$

   b) **a** is 5 km North and **b** is 2.5 km West

   c) $\mathbf{a} = 3\mathbf{i} + 2\mathbf{j}$ and $\mathbf{b} = -7\mathbf{i} + 5\mathbf{j}$

2. 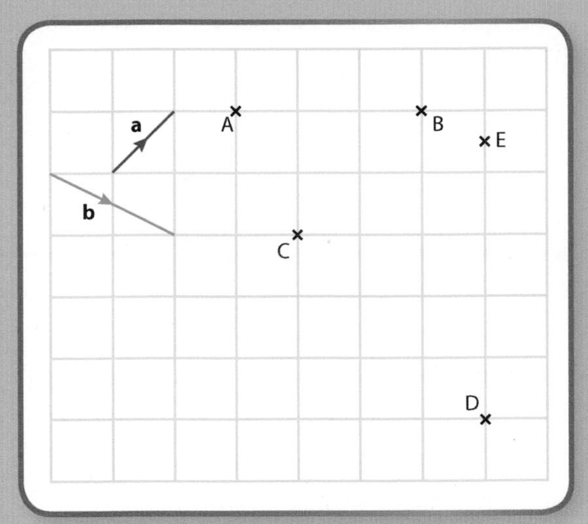

   Two vectors have been defined on the grid. Use the vectors to describe these:

   a) $\overrightarrow{AB}$ b) $\overrightarrow{BC}$ c) $\overrightarrow{BD}$ d) $\overrightarrow{DE}$

   e) A point F is collinear with DE. Write a general expression for the vector $\overrightarrow{EF}$.

   f) A point G is such that the vector $\overrightarrow{EG}$ is parallel to $\overrightarrow{AC}$ but has twice the magnitude. What is the vector $\overrightarrow{EG}$?

3. The points A and B have position vectors $\overrightarrow{OA} = 12\mathbf{i} - 7\mathbf{j}$ and $\overrightarrow{OB} = -3\mathbf{i} + 7\mathbf{j}$.

   a) Find the vector $\overrightarrow{BA}$.

   b) State the vector $\overrightarrow{AB}$.

PRACTICE QUESTIONS

1. A trapezium ABCD has parallel sides AD and BC, where AD is three times the length of BC. The vectors $\mathbf{p} = \overrightarrow{AD}$ and $\mathbf{q} = \overrightarrow{DC}$ are shown on the diagram.

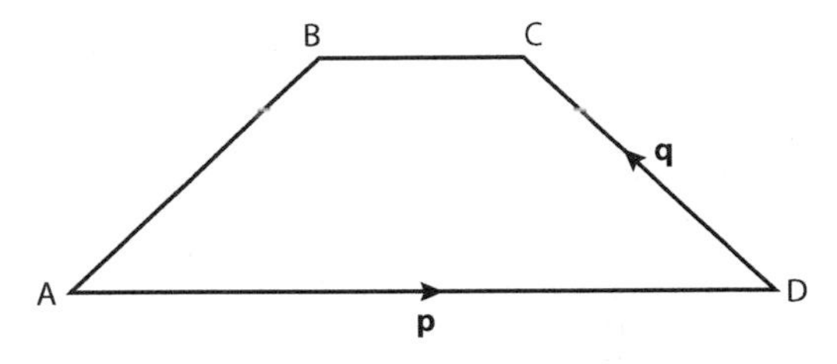

   a) Find the vector $\overrightarrow{AB}$ in terms of **p** and **q**. **[2 marks]**

   b) Given that point E is collinear with D and C, find an expression for $\overrightarrow{AE}$. **[2 marks]**

   c) A point M lies on AD such that AM : MD = 6 : 1. Given that E is also collinear with points B and M, find an expression for $\overrightarrow{BE}$ in terms of **p** and **q**. **[7 marks]**

2. From a lighthouse O, points A and B have position vectors $\overrightarrow{OA} = q\mathbf{i} + 12\mathbf{j}$ km and $\overrightarrow{OB} = p\mathbf{i}$ km, where p and q are positive integer values.

   a) Find, in terms of p and q, the vector $\overrightarrow{AB}$. **[2 marks]**

   b) A ship starting at point A, which is 37 km away from the lighthouse, sails on a bearing of 202.6° (1 d.p.) to reach B. What is the distance AB? **[5 marks]**

Kinematics 2

Equations of Constant Acceleration

When considering motion in a straight line with constant acceleration, you can use the *suvat* equations, so named after the variables:

- s = displacement
- u = initial velocity (at the start of the time period being considered)
- v = final velocity (at the end of the time period being considered)
- a = acceleration
- t = time (this is the duration of the motion rather than the time at which it happens)

As vector quantities, the positive direction needs to be defined and applied consistently to all the values. Take note of the direction in interpretations of answers.

Deriving the Equations of Constant Acceleration

You must be able to remember and recognise the equations for constant acceleration, as well as be able to derive them from given assumptions. A classic starting point from which to derive the equations is to use a velocity–time (v–t) graph. Since the acceleration is constant, the graph is linear.

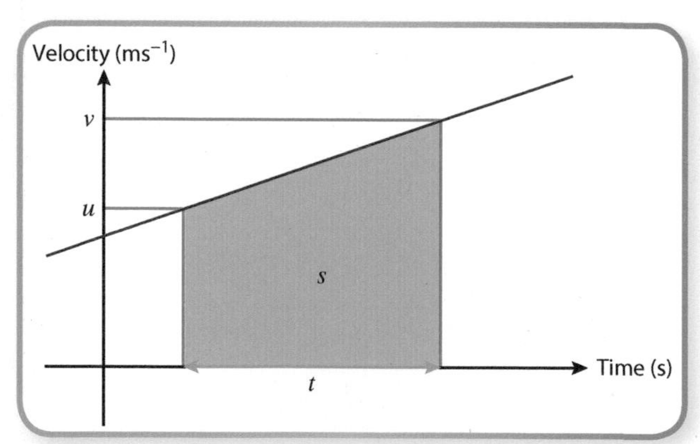

The **area** under a v–t graph is the displacement. To find the area, use the area of a trapezium to give:

$$\boldsymbol{s = \tfrac{1}{2}(u + v)t}$$

This formula is used when the acceleration isn't given or asked for.

The **gradient** of the line is the acceleration. To find the gradient, $\frac{\text{change in velocity}}{\text{time for change to occur}}$:

$a = \frac{v-u}{t}$, which rearranges to give: $\boldsymbol{v = u + at}$

This formula is used when the displacement isn't given or asked for.

Combining the two previous equations (substituting the second into the first) gives:

$$s = \tfrac{1}{2}(u + u + at)t$$

$$\boldsymbol{s = ut + \tfrac{1}{2}at^2}$$

This formula is used when the final velocity isn't given or asked for.

In a similar way, to create an equation without u, rearrange $v = u + at$ to give $u = v - at$, then substitute into $s = \frac{1}{2}(u + v)t$ to give:

$$s = \tfrac{1}{2}(v - at + v)t$$

$$\boldsymbol{s = vt - \tfrac{1}{2}at^2}$$

This formula is used when the initial velocity isn't given or asked for.

So the only variable left to be eliminated from a formula is time. By rearranging $v = u + at$ to give $t = \frac{v-u}{a}$, then substituting into $s = \frac{1}{2}(u + v)t$, gives:

$$s = \tfrac{1}{2}(u + v)\left(\frac{v-u}{a}\right)$$

$$2as = (u + v)(v - u)$$

$$2as = v^2 - u^2$$

$$\boldsymbol{v^2 = u^2 + 2as}$$

This formula is used when the time isn't given or asked for.

Using the Equations of Constant Acceleration

The equations can be used to find any one of the variables (s, u, v, a or t) within the equation. The skill is in knowing/deriving the best equation(s), interpreting the context correctly and identifying all the known values accurately (including negatives/positives).

Questions can be split into sections, representing different accelerations. Explain your work clearly and do not get confused between different *suvats*.

Example

A helicopter is flying over the sea at a constant height in a straight line. It is accelerating at a constant $0.2\,\text{ms}^{-2}$ towards a lighthouse. At $t = 0$ it is travelling at a speed of $1\,\text{ms}^{-1}$ away from the lighthouse. The lighthouse is 4 km away from the helicopter's starting position. Find the velocity of the helicopter when it is vertically above the lighthouse.

Start by identifying each unknown and whether it is a given, needs to be found or is irrelevant. Check the consistency of units at this point.

Define the positive and negative direction. This can be either way as long as all values are used and interpreted correctly in the system.

Positive direction is from the initial position of the helicopter towards the lighthouse. This could be shown on a diagram.

$s = 4000\,\text{m}$, $u = -1\,\text{ms}^{-1}$, $v = ?$, $a = 0.2\,\text{ms}^{-2}$, t = irrelevant

You can now select the correct equation. Write down the equation that is going to be used.

$$v^2 = u^2 + 2as$$

$$v^2 = (-1)^2 + 2 \times 0.2 \times 4000 = 1601$$

$$v = \sqrt{1601} = 40.0124\ldots$$

$$= 40.0\,\text{ms}^{-1} \text{ (3 s.f.)}$$

Calculus and Kinematics

$v = \frac{dr}{dt}$ Velocity is the rate of change of position with respect to time.

$r = \int v\,dt$ Position is the integral of the velocity with respect to time.

$a = \frac{dv}{dt} = \frac{d^2r}{dt^2}$ Acceleration is the rate of change of velocity with respect to time.

$v = \int a\,dt$ Velocity is the integral of the acceleration with respect to time.

Kinematics provides a context for calculus skills. Questions may not specify that calculus is required but may give an equation in terms of t for r, v or a.

If acceleration is constant then graphs or equations of constant acceleration might be the better option. If the equation for velocity isn't linear (i.e. acceleration is variable) then calculus will be helpful. An exam question may require multiple techniques to solve it, or a second method may provide a neat way of checking an answer.

Example

A car moving in a straight line has a velocity of $v = -2t^2 + 4t + 6$ $(t \geqslant 0)$.

a) For what values of t does the car have positive acceleration?

$$a = \frac{dv}{dt} = -4t + 4 \qquad -4t + 4 > 0$$

$$4t < 4$$

$$0 \leqslant t < 1$$

b) Find the distance travelled between $t = 1$ and $t = 3$, correct to 2 decimal places.

$$r = \int_1^3 -2t^2 + 4t + 6\ dt$$

$$= \left[-\tfrac{2}{3}t^3 + 2t^2 + 6t\right]_1^3$$

$$= \left(-\tfrac{2}{3} \times 3^3 + 2 \times 3^2 + 6 \times 3\right) - \left(-\tfrac{2}{3} \times 1^3 + 2 \times 1^2 + 6 \times 1\right)$$

$$= (18) - \left(\tfrac{22}{3}\right)$$

$$= \frac{32}{3} = 10.67 \text{ (2 d.p.)}$$

c) When $t = 1$ the car is at the origin. Find an expression for the displacement from the origin of the car at time t.

$$r = \int v\,dt = -\tfrac{2}{3}t^3 + 2t^2 + 6t + C$$

$$0 = -\tfrac{2}{3} + 2 + 6 + C$$

$$C = \tfrac{2}{3} - 2 - 6 = -\tfrac{22}{3}$$

$$r = -\tfrac{2}{3}t^3 + 2t^2 + 6t - \tfrac{22}{3}$$

Acceleration Due to Gravity

A classic example of motion with constant acceleration is that of objects being dropped or projected. Modelling assumptions are key:

- The object is modelled as a **particle** – this means there is no **air resistance** (if air resistance is included, acceleration would vary with speed). Sometimes this is a reasonable model as the impact of air resistance on, for example, a snooker ball as it falls off a table would have very little effect. At other times this isn't realistic, e.g. a person skydiving will hit a terminal velocity in freefall when their weight is balanced by the air resistance. The effect over a smaller distance, such as a person jumping off a diving board, would be less significant.
- Being a particle also means that the object is considered not to spin and has no size. Again, consider the context as to whether this is reasonable or not. The errors incurred by such assumptions may be negligible (e.g. if dropping a marble off a tall building) or significant (e.g. if dropping a football 1 m).
- **Projectiles** can be launched and land at the same level. The assumption is that the launcher has no vertical height above the 'ground'. Questions sometimes use different heights (e.g. a diver jumping into a pool from a high board) but if something is projected at an angle to the 'ground' it is assumed that the height of release and landing can be the same. Again, if asked to assess the impact of this modelling assumption, consider the size of the simplification compared to the overall distances involved.

Objects are considered to be 'launched', giving them an initial velocity. This velocity can be split into two components, horizontal and vertical.

Once released, the object experiences no further forces horizontally so has a constant horizontal velocity. Vertically it experiences acceleration due to gravity g, which is ~9.8 ms^{-2}. Questions may require a different level of accuracy to be taken for g (e.g.10). If not specified, it is worth working in terms of g for as long as possible, especially as terms often cancel out.

Be clear about what direction is positive. If upwards is positive, then $g = -9.8\text{ ms}^{-2}$.

Example

A skydiver jumps from a plane and whilst in freefall can be modelled as a particle. She jumps up and out with a horizontal velocity of 2 ms^{-1} and with the upwards component of her velocity being 2 ms^{-1}. Use $g = 9.8\text{ ms}^{-2}$.

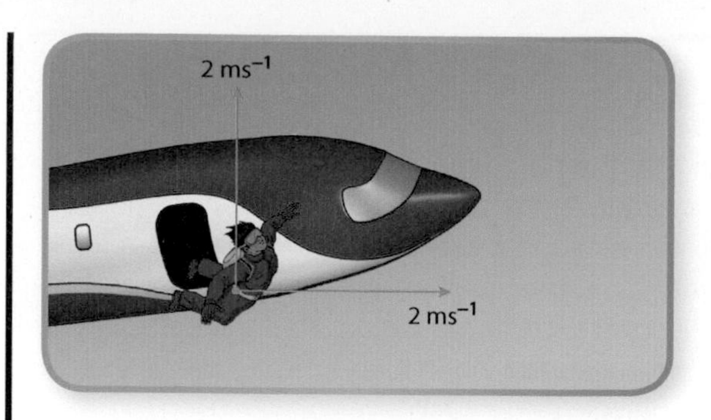

a) Find the horizontal displacement at the point when the skydiver has a vertical displacement of 378.9 m.

Horizontal displacement $= vt = 2t$

Vertically (positive is downwards):

$s = 378.9$, $u = -2$, $v =$ irrelevant, $a = 9.8$, $t = ?$

$378.9 = -2t + \frac{1}{2} \times 9.8 \times t^2$

$3789 = -20t + 49t^2$

$49t^2 - 20t - 3789 = 0$

$t = \frac{20 \pm \sqrt{(-20)^2 - 4 \times 49 \times (-3789)}}{98}$

$t = 9 \quad (t \neq -8.591\ldots$ in this context)

Substitute into horizontal:

Displacement $2 \times 9 = 18$ m

b) Comment on the modelling assumptions and the expected effect they would have compared to real-life measurements of the situation.

Even if the skydiver tucked up into a 'ball', she would still experience air resistance. This would reduce the acceleration so the time taken to fall a set distance would be more than that calculated.

The skydiver could also be affected by wind conditions. This could mean that the horizontal velocity isn't constant and could affect the horizontal distance travelled.

A skydiver can control their velocity to some extent (by changing their body shape), so the modelling assumption could be refined to include the variation in acceleration due to air resistance. If more information was provided about wind conditions and body position, a better model could be used.

SUMMARY

- **Equations of constant acceleration/equations of constant motion/*suvat* equations:**

  $s = \frac{1}{2}(u+v)t$ $\quad$ $s = ut + \frac{1}{2}at^2$

  $v = u + at$ $\quad$ $s = vt - \frac{1}{2}at^2$

  $v^2 = u^2 + 2as$

- **Calculus applied to kinematics:**

  $v = \frac{dr}{dt}$ $\quad$ $a = \frac{dv}{dt} = \frac{d^2r}{dt^2}$

  $r = \int v \, dt$ $\quad$ $v = \int a \, dt$

- **Check positivity/negativity.**
- **Remember $+C$ when integrating (unless between bounds).**
- **A quadratic, for example when finding t using $s = ut + \frac{1}{2}at^2$, will give two answers. Acknowledge both even if one can be discounted.**
- **Make sure the units are consistent. It isn't necessary to convert units to use the equations, but all the units must have the same measure of time and distance.**

QUICK TEST

1. Without looking at notes, write down the equation of constant motion (*suvat* equation) that you would use in the given situation:
   a) Given s, u, v and asked to find a
   b) Given a, v, t and asked to find u
   c) Given a, u, s and asked to find t
   d) Given v, t, a and asked to find s
   e) Given t, u, s and asked to find v
2. Find ? in these (all values are in SI units).
   a) $s = 37.5$, $u = ?$, $v = 7$, $a =$ irrel., $t = 15$
   b) $s =$ irrel., $u = -12$, $v = 8$, $a = ?$, $t = 20$
   c) $s = ?$, $u = 7$, $v = 17$, $a = 5$, $t =$ irrel.
3. Given that $r = \frac{1}{2}t^4 - 2t^2 - 15$
   a) Find an expression for v in terms of t.
   b) Find the value of v when $t = 2$.
4. A motorist is driving at $26\,\text{ms}^{-1}$. He spots a hazard 100 m in front and brakes with constant deceleration, coming to rest 9 m before the obstacle.
   a) How long does it take the car to come to rest?
   b) What is the deceleration of the car?

PRACTICE QUESTIONS

1. Given that $a = 3t + 2$
   a) Find an expression for v in terms of t. **[2 marks]**
   b) Find an expression for r in terms of t. **[3 marks]**
   c) Given that at $t = 1$, $r = 3.5$, and at $t = 2$, $r = 13$, find the value of the displacement at $t = 3$. **[5 marks]**
2. A marble is rolling along a horizontal table with an initial velocity of $6\,\text{cms}^{-1}$. It is decelerating at $0.2\,\text{cms}^{-2}$ due to constant friction along the table. At $t = 0$ the marble is half a metre from the edge of the table. Use $g = 9.8\,\text{ms}^{-2}$.
   a) Show that the marble wlll reach the edge of the table. **[3 marks]**
   b) The marble rolls off the edge of the table and falls freely to the floor. The table is $\frac{3}{4}$ m tall. Find the horizontal distance from the edge of the table to where the marble hits the floor. **[6 marks]**
   c) State and explain a modelling assumption used to answer parts **a)** and **b)**. **[1 mark]**

Links to Other Concepts
- Solving quadratics
- Integration and differentiation
- Interpreting graphs
- Solving simultaneous equations
- Vectors
- Algebraic manipulation
- Interpreting context-based questions

Forces

Forces are measured in Newtons (N). 1 Newton is the force required to accelerate a mass of 1 kg at 1 ms^{-2}.

Types of Force

A force is generally a 'push' or a 'pull'. The forces met at AS-level are:

- **Normal reaction** – a balancing force created by a surface when an object is pressed into it. For example, a person standing on the floor exerts a downwards force (weight) onto the surface. The surface is non-active and it doesn't do any pushing; it just resists the force applied to it and, in doing so, an equal and opposite force is created. Importantly the normal reaction force is always at **right angles to the surface**. The normal reaction force is equal to a component of the weight which is in the same direction. A normal reaction force could be horizontal if the surface is vertical and the object is being pushed into it.
- **Weight** – the force of attraction between two bodies with mass. Weight is the mass of an object (the amount of material it is made of) multiplied by the acceleration due to gravity, i.e. $W = mg$. Mass and weight are different. Mass is not a force. An object still has mass whilst in space but without gravity it would have no weight. In most cases the acceleration due to gravity is taken as ~ 9.8 ms^{-2}. A question could be set with a different value of g, say on the Moon, but this will be made clear in the question.
- **Tension** – a pulling force created within a 'string' or 'rod'. It is created when the rod is in tension (being pulled from each end). The rod transfers the force through it, which is experienced as a pull along the rod. The pull is experienced equally, but in opposite directions, at each end of the rod.
- **Thrust** – a pushing force. This is a key difference between something modelled as a string (no thrust can be transferred) and a rod (thrust is transferred). As with tension, the force is transferred through the rod. The thrust from a rod is experienced equally, but in opposite directions, at each end of the rod.

Example

Explain why tow bars are preferable when towing a car to the traditional use of ropes.

For the majority of the time the rope/rod is in tension, so both rope and rod will work to pull the car along. However, when the front vehicle needs to brake (or goes down a hill), the rod will transfer the braking force to the car being towed. The rope doesn't do this as it cannot transfer a thrust force. This means, if ropes are used, the driver in the rear car has to be ready to brake when the rope goes slack.

- **Friction** – a force that occurs between a rough surface and an object that are in contact. It acts along the surface in the direction opposite to motion. If an object is held in equilibrium (all forces are balanced) then friction can be a balancing force up to a maximum value of μR, where μ is the coefficient of friction between the object and the surface and R is the normal reaction force between the object and the surface. When the object is moving, friction is equal to μR.

Force Diagrams

Forces are often represented in force diagrams. As modelling at AS-level represents bodies as particles, diagrams show all the forces coming out from the point which represents the body in question. Force is a vector quantity and, as such, the lines show a magnitude and a direction. They can be represented as any vector, using **i** and **j** notation or column vectors, or by specifying a magnitude and a direction.

Equilibrium

Forces are in equilibrium when they are balanced, i.e. there is no resultant force. When forces are in equilibrium, there is no acceleration experienced by an object so it is either stationary or continues to move with a constant velocity (this is Newton's first law of motion).

Example

Three forces are acting on a particle at right angles to each other, as shown in the diagram. A fourth force is added such that the particle is in equilibrium.

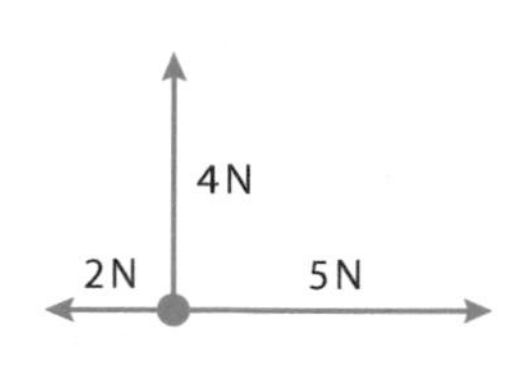

a) Show in a diagram the sum of the forces and the fourth force needed to create equilibrium.

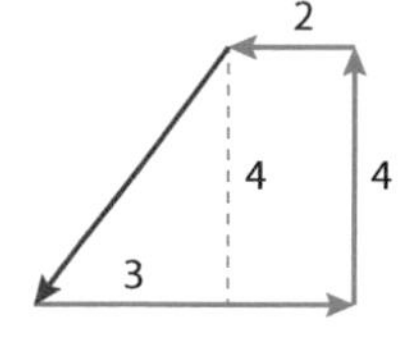

b) Hence find the magnitude and the direction of the new force.

Magnitude $= \sqrt{3^2 + 4^2} = 5$

Direction is $\tan^{-1}\left(\frac{3}{4}\right) = 36.9°$ (1 d.p.) from the downwards vertical.

Using a component vector approach allows two equations to be created showing a balance, or imbalance, of the forces in two directions acting at right angles to each other.

Example

Four forces, A, B, C and D, are applied to a particle on a smooth horizontal plane.

$A = \begin{pmatrix} 2 \\ p+1 \end{pmatrix}$, $B = \begin{pmatrix} p \\ -7 \end{pmatrix}$, $C = \begin{pmatrix} 2 \\ 0 \end{pmatrix}$ and $D = \begin{pmatrix} -6 \\ q \end{pmatrix}$

Given that the particle is moving with a constant velocity of 3 ms^{-1}, find the values of p and q.

As all forces are split into components, you can equate the separate components at a time. The forces must balance in the y-direction and the x-direction to be in equilibrium. The particle is in equilibrium so the sum of the forces equals zero.

'x-direction': $2 + p + 2 - 6 = 0$
$-2 + p = 0$
$p = 2$

'y-direction': $(p+1) - 7 + 0 + q = 0$
$3 - 7 + q = 0$
$q = 4$

Friction, Normal Reaction and Equilibrium

Normal reaction is the force that creates equilibrium at right angles to the surface that an object is on.

$F_r = \mu R$ only applies when the object is at the point of slipping or in motion. If the body is stationary, then $F_r \leqslant \mu R$. It can act either way along a plane but is opposite to the direction of motion, or potential motion if 'at the point of slipping'.

Example

A box is sitting on a slope, as shown in the diagram. It is kept in equilibrium, but on the point of slipping down the plane, by a rope that acts along the slope.

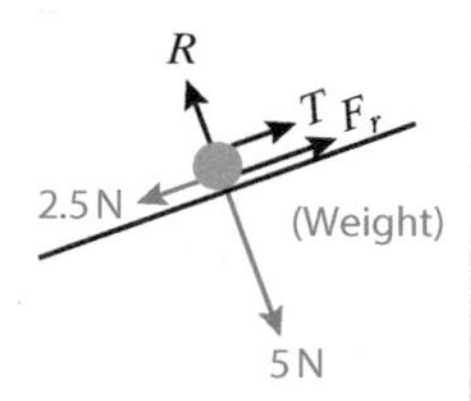

The weight of the box has been split into two forces, one acting along the plane and the other acting at right angles to the plane.

a) What is the value of the normal reaction force between the plane and the box?

Components perpendicular to the slope are in equilibrium and the resultant force is 0.

$R - 5 = 0$, so the normal reaction = 5 N

b) Find the coefficient of friction between the box and the plane given that the tension in the rope is 0.5 N.

$F_r = 2.5 - T = 2$ N

Point of slipping so $F_r = \mu R$

$2 = 5\mu$

$\mu = \frac{2}{5} = 0.4$ (Note: μ doesn't have a unit.)

Resultant Forces

A resultant force is the sum of all the forces acting on a body.

Example

The force diagram shows five forces acting on a body. Find the magnitude and the direction, given as a bearing, of the resultant force.

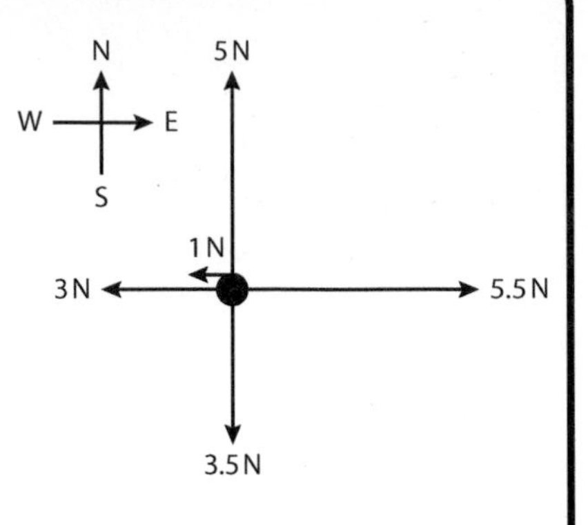

Vertical component of the resultant force
$= 5 - 3.5 = 1.5$ to the North

Horizontal component of the resultant force
$= 5.5 - 3 - 1 = 1.5$ to the East

Overall resultant force:

Magnitude $= \sqrt{1.5^2 + 1.5^2} = \frac{3\sqrt{2}}{2}$

At a bearing of 045°

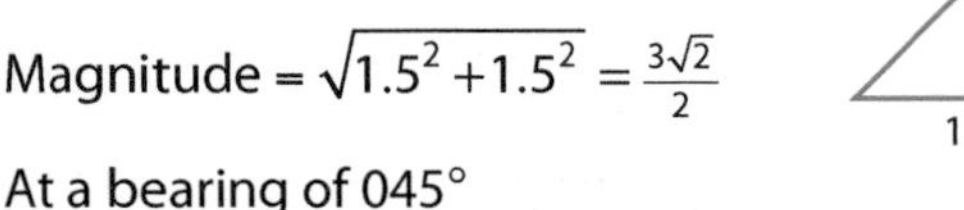

Newton's Second Law of Motion, $F = ma$

Newton's second law states that the resultant force (F) is equal to the mass multiplied by the acceleration of the object. This can be rearranged to find m or a:

$m = \frac{F}{a}$ $\quad a = \frac{F}{m}$

Example

Louise is being lowered on a winch from a hovering helicopter. She weighs 85 kg, and experiences no external forces other than the tension in the cable. The cable provides a constant tension T. Given that Louise accelerates at a rate of $2\,\text{ms}^{-2}$ during the first phase of motion, find the value of the tension in the cable. The winch cable is modelled as light and inextensible. Louise is modelled as a particle. Take $g = 9.8\,\text{ms}^{-2}$.

T

$85g$

$85g - T =$ resultant force
$F = ma = 85 \times 2 = 170$
$85 \times 9.8 - T = 170$
$T = 85 \times 9.8 - 170$
$= 663\,\text{N}$

Newton's Second Law of Motion with Equations of Constant Acceleration

Example

Louise is lowered from the helicopter starting with 0 velocity, and after 4 seconds the tension in the winch cable is increased to 1003 N, which allows her to land on the ground travelling at $4\,\text{ms}^{-1}$. Find the height at which the helicopter is hovering.

Plan for answering question:

1. *Use first period, and suvat, to find final velocity.*
2. *Use first period to find displacement.*
3. *Use $F = ma$ to find new acceleration for second period of motion.*
4. *Use velocity from* **1** *and acceleration from* **3** *to find displacement.*
5. *Add the displacements together.*

For first period of motion:
$s = ?, u = 0, v = ?, a = 2, t = 4$

(Note: Positive direction is being taken as downwards)

To find v, use $v = u + at$
$v = 0 + 2 \times 4 = 8\,\text{ms}^{-1}$

To find s use any equation, but if possible use the values given by the question rather than calculated:

$s = ut + \frac{1}{2}at^2$

$s = 0 + \frac{1}{2} \times 2 \times 4^2 = 16\,\text{m}$

For second period of motion:

$F = ma$
$85g - 1003 = 85a$
$a = \frac{85g - 1003}{85} = -2\,\text{ms}^{-2}$

(Note: The negative implies upwards acceleration)

$s = ?, u = 8, v = 4, a = -2, t =$ irrel.

To find s, use $v^2 = u^2 + 2as$

$4^2 = 8^2 + 2 \times -2 \times s$

$s = \frac{16 - 64}{-4} = 12\text{m}$

Total distance from the hovering helicopter to the ground $= 12 + 16 = 28\,\text{m}$

Questions will not always specify the method(s) to be used, so always consider what information is being provided and what is asked for. If acceleration is constant then the equations of constant motion are a good starting point. If there are multiple periods of motion, or imagining the situation is challenging, a velocity–time or displacement–time graph might help.

Connected Particles

Connected particles can be modelled as two bodies separately. The tension is equal through the string/rod/rope so is felt equally, but possibly in different directions, by each object. Acceleration is also constant assuming the rope stays taut. This often leads to a pair of simultaneous equations to solve in terms of a and T.

Horizontal

A car that pulls a trailer can be modelled as two separate bodies and force diagrams drawn for each. It isn't significant what is pulling the trailer – only the force that is exerted matters – so this is what is represented by the diagram.

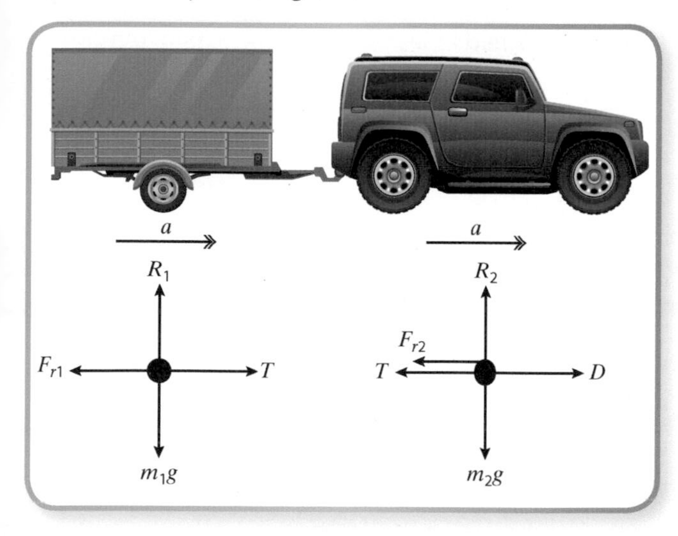

T is the same magnitude in each diagram. The acceleration is also the same for the car and the trailer. This is shown by the double-headed arrows.

Vertically the vehicle and trailer are in equilibrium – they are travelling along the road. The vertical forces are significant though, as they determine the different frictional forces F_{r1} and F_{r2}. It is also possible to model the car and trailer as a single body, as they are travelling together with the same velocity:

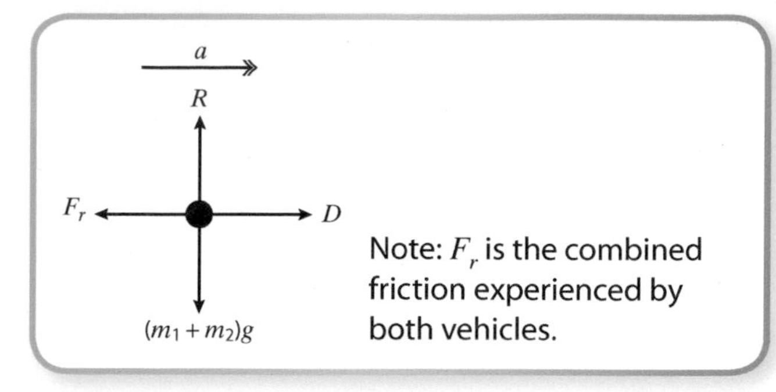

Note: F_r is the combined friction experienced by both vehicles.

Vertical

Two particles are connected by a string that passes over a smooth pulley (this could be a lift and its counterweight).

Again each body can be represented separately. The pulley adds the fact that the acceleration and the motion, whilst still equal in magnitude, are opposite in direction.

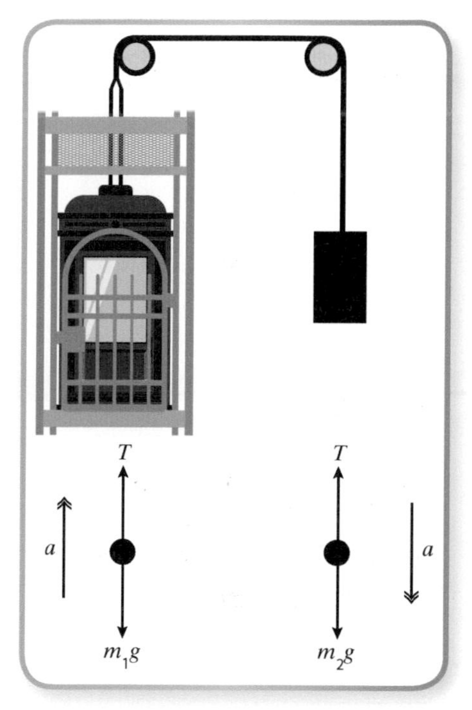

If an extra equation is needed to help to solve the problem then you can consider it as one body, but first the directions need to be aligned so that the acceleration is in the same direction.

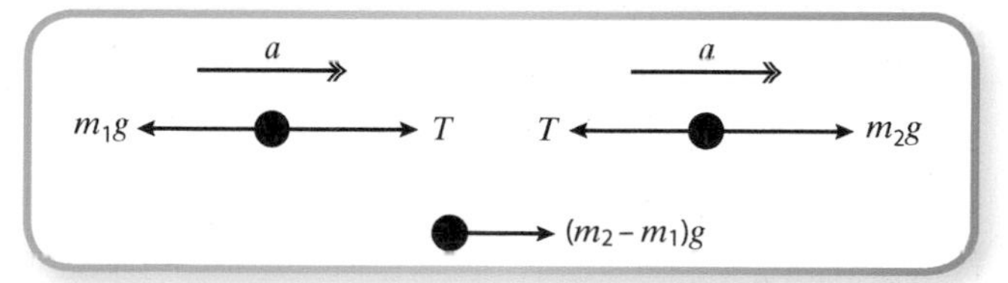

Don't forget

By modelling the two objects as one body, the tension force is eliminated. Light, inextensible string means the two bodies will have the same acceleration and velocity at all times (as long as the string is taut).

DAY 6

It is also possible to have one object travelling horizontally whilst another is travelling vertically.

Example

A child has tied together two toys with a length of string. She places them both on a table, which is 58 cm tall.

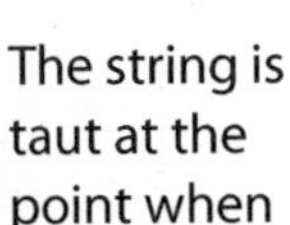

The string is taut at the point when the tractor, which has a mass of 270 g, falls off the edge of the table. The teddy, whose mass is 180 g, is pulled along the rough horizontal surface of the table top. Given that $\mu = 0.5$:

a) Find the acceleration of the teddy in terms of g.

Force diagrams for each object:

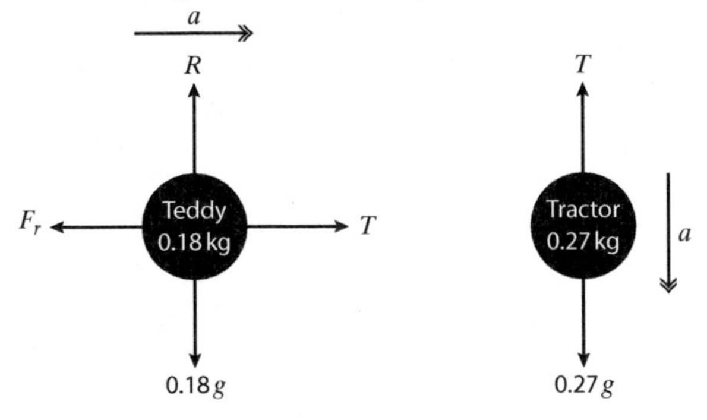

Tractor: $F = ma$

Resultant force $F = 0.27g - T$

$0.27a = 0.27g - T$

Teddy vertically in equilibrium: $R = 0.18g$

Teddy horizontally: $F = ma$

Resultant force $F = T - \mu \times 0.18g = T - 0.09g$

$0.18a = T - 0.09g$

Solve as simultaneous equations (finding both unknowns adds a check in the working):

$T = 0.18a + 0.09g$ (rearrangement of horizontal equation for teddy)

$0.27a = 0.27g - (0.18a + 0.09g)$ (substituted into tractor vertical equation)

$0.27a = 0.18g - 0.18a$

$0.45a = 0.18g$

$a = 0.4g\,\text{ms}^{-2}$

b) Find the velocity of the tractor as it hits the ground. Use $g = 10\,\text{ms}^{-2}$.

Using equations of constant acceleration:

$s = 0.58$, $u = 0$, $v = ?$, $a = 4$, $t = $ irrelevant

$v^2 = u^2 + 2as$

$v^2 = 0 + 2 \times 4 \times 0.58 = 4.64$

$v = \sqrt{4.64} = 2.15\,\text{ms}^{-1}$ (3 s.f.)

c) State a modelling assumption you have made in answering this question.

The toys have been modelled as particles, so they are a point mass and aren't affected by air resistance.

The string is modelled as light and inextensible, so the tension is transferred through the string. The objects are travelling with the same speed and magnitude of acceleration.

The string passing over the edge of the table is modelled as a smooth pulley. Again, tension is transferred fully through the string.

SUMMARY

- **Forces are vectors and can be represented using force diagrams.**
- **An object is in equilibrium when the resultant force is 0.**
- **An object in equilibrium will have zero acceleration – this means it will maintain its velocity (which could be stationary).**
- **If there is a resultant force acting on a body, it will accelerate in the direction of the force.**
- **$F = ma$ (Newton's second law of motion)**
- **Connected particles have the same velocity and acceleration (assuming the connection is taut throughout the motion).**
- **The tension through the connection is of equal magnitude.**
- **Bodies that are connected can be modelled as a single body or considered separately.**

QUICK TEST

1. Find the resultant force in these diagrams.

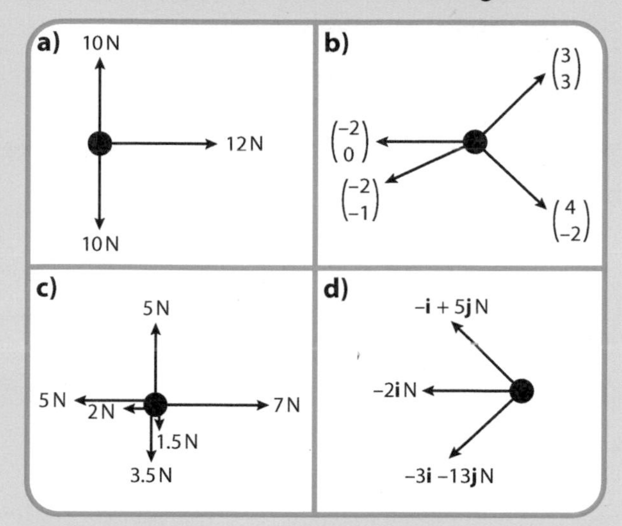

2. For each diagram above, find the acceleration experienced by the object, given that they have the following masses:

   a) 3 kg **b)** 40 g **c)** 7 kg **d)** 500 g

3. Draw a force diagram representing the forces acting on this wakeboarder.

4. A car is pulling a caravan along a straight, horizontal road. The car is producing a driving force of 180 N. The car has mass 1500 kg and the caravan 750 kg. Both vehicles experience constant resistance forces (including friction) of 15 N.

   a) Draw a force diagram to represent both the car and the caravan as a single body.

   b) Find the acceleration of the caravan.

   c) Draw separate force diagrams for the car and the caravan.

   d) Find the tension in the tow bar.

PRACTICE QUESTIONS

1. A car, of mass 1412 kg, is pulling a trailer, of mass 610 kg. The car experiences a constant resistance to motion of 340 N and the trailer a constant resistance of 210 N.

   a) Find the driving force produced by the car to create an acceleration of $2\,\text{ms}^{-2}$. **[3 marks]**

   Sometime later the car is producing a driving force of 2976.4 N (resistive forces unchanged). It passes point A, on a straight and horizontal road, travelling at $12\,\text{ms}^{-1}$. 10 seconds later, it passes point B where the tow bar disconnects.

   b) The trailer comes to a rest at point C. What is the distance BC? **[8 marks]**

   c) The car continues with the same driving force to point P, a quarter of the distance from B to C. What constant braking force would the driver need to apply at P in order to come to a rest at C? **[7 marks]**

2. A lift has mass 1200 kg and can take a load of up to 400 kg. The lift uses a counterweight of 1400 kg.

   Modelling the lift and the counterweight as particles connected by a light, inextensible string passing over a smooth pulley:

   a) Find the maximum acceleration possible in the lift, in terms of g, when there is no driving force applied. **[7 marks]**

   b) What would be the tension in the cable for the above situation? **[2 marks]**

Links to Other Concepts

- Solving quadratics
- Integration and differentiation
- Interpreting graphs
- Solving simultaneous equations
- Vectors
- Algebraic manipulation
- Interpreting context-based questions

Proof

All mathematics is based on a small set of rules from which everything has been proved to be true. Proof at AS-level doesn't go all the way back to the basic axioms of mathematics – for example, it is fine to use arithmetic without proving from first principles. It is important to have a firm grasp of types of number and the ability to show detailed and well-explained steps in working.

Key Concepts and Notation

Set Notation and Types of Number

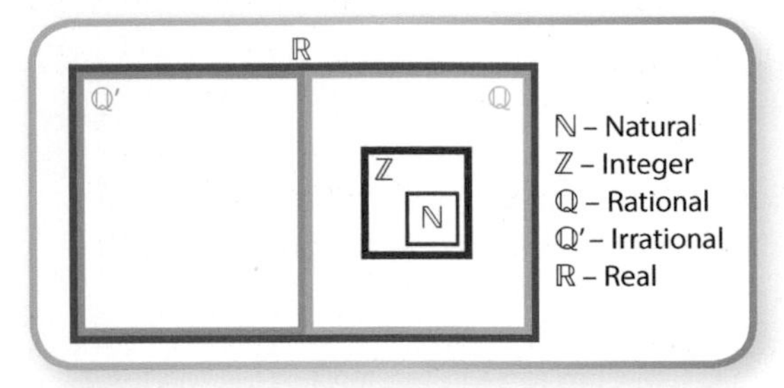

$\mathbb{N}$ – Natural numbers; they are the positive integers and zero. 0, 1, 2, 3, 4, 5, 6, …

$\mathbb{Z}$ – Integers/whole numbers; these include the negatives. …, –4, –3, –2, –1, 0, 1, 2, 3, …

$\mathbb{Q}$ – Rational; these are numbers that can be written accurately as a fraction. They include the integers but also include any fraction or decimal with recurring decimal places.

$\mathbb{Q}'$ – Irrational; numbers that cannot be written accurately as a fraction, e.g. $\sqrt{2}$, π and e.

$\mathbb{R}$ – Real numbers; all the numbers mentioned so far. They are the numbers that would fit somewhere on a traditional number line.

The groups can be subdivided into positives and negatives using superscript, e.g. $\mathbb{Z}^-$ would be the negative integers. Just as $\mathbb{Q}'$ is 'not' rational, $\mathbb{N}'$ is 'not' natural.

The symbol $\in$ means 'is an element of'. It is used with a variable and a set of numbers. For example, $x \in \mathbb{N}$ means x belongs to the group of natural numbers. It is an element within this set.

You can use the element symbol with a defined set of numbers, e.g. $n \in \{2, 3, 6, 8, 11\}$ means that n is an element of the set of numbers defined within the brackets.

$\Rightarrow$ means 'implies'.

Factors, Multiples, Even Numbers and Odd Numbers

Factors and multiples deal with natural numbers and the ability to break down numbers and express them as a multiplication of their factors.

Working with integers, a number is considered to be even if it can be split into two equal-sized groups (where each group is an integer value). In other words, an even number has a factor of 2. Odd numbers don't have a factor of 2.

Prime Numbers

A prime number has only two factors: 1 and itself. You must be able to recognise the prime numbers up to about 50, and have a method for establishing whether bigger numbers are prime or not (primality).

Prime numbers up to 100: **2, 3, 5, 7, 11, 13, 17, 19, 23, 29, 31, 37, 41, 43, 47, 53, 59, 61, 67, 71, 73, 79, 83, 89, 97**

If presented with a number that is not instantly recognisable as prime or non-prime, you can test primality by dividing the number by the primes up to the square root of the number.

Example

Determine if 299 and 347 are prime.

By inspection there are no obvious small factors for either value. Quick checks:

2 – the numbers are not even so 2 is not a factor (even numbers end in a 0, 2, 4, 6 or 8)

3 – the sum of the digits isn't divisible by 3 so neither number is divisible by 3

5 – the numbers do not end in a 0 or a 5 so 5 is not a factor

You could carry out these calculations on a calculator. If it gives an integer value, the number is not prime.

$\sqrt{299} = 17.291\ldots$ so try dividing by the primes up to 17:

$299 \div 7 = \frac{299}{7} = 42.7142\ldots$ $\therefore$ 7 is not a factor

$299 \div 11 = \frac{299}{11} = 27.181\ldots$ $\therefore$ 11 is not a factor

$299 \div 13 = 23$ $\therefore$ 13 is a factor, as is 23

$\therefore$ 299 is not prime.

$\sqrt{347} = 18.627\ldots$ so try dividing by the primes up to 18:

$347 \div 7 = \frac{347}{7} = 49.571\ldots$ $\therefore$ 7 is not a factor

$347 \div 11 = \frac{347}{11} = 31.545\ldots$ $\therefore$ 11 is not a factor

$347 \div 13 = \frac{347}{13} = 26.692\ldots$ $\therefore$ 13 is not a factor

$347 \div 17 = \frac{347}{17} = 20.411\ldots$ $\therefore$ 17 is not a factor

$\therefore$ 347 is a prime number.

The above demonstrates two proofs: the first by counter-example and the second by exhaustion. If asked to prove that either was a prime number, more detail would be included.

Types of Proof

A proof question will give a starting point and a final destination and it will require all the intermediate directions to be given. Any question that asks 'show that' is a form of proof.

Proof by Deduction (Direct Proof)

Proof by deduction is based on the idea that from an assumed statement it is possible to deduce a series of other statements until the proof is achieved. Make sure each statement supports the next step.

Deduction often uses a generalisation, e.g. let the number = n, where n is any integer value. Then if the rule can be shown to hold true for n, it is true for all integer values and so has been proved.

Commonly-used generalisations include:

- n, representing any number, often defined as $n \in \mathbb{Z}$, as with nth terms.
- Then $2n$ could be used to represent any even number, and $2n + 1$ any odd number.
- A pair of consecutive integers would be n and $n + 1$ (or $n - 1$ and n).
- n^2 would represent any square number.

If a second generalised number is needed, m is often used. However, any clearly defined letter is acceptable.

Example

Here are positive integers organised into a table which is five columns wide. Prove that the sum of any two-by-two square of numbers, taken from within the table, is a multiple of 4.

1	2	3	4	5
6	7	8	9	10
11	12	13	14	15
16	17	18	19	20
21	22	23	24	25
26	27	28	29	30
31	32	33	34	35
36	37	38	39	40
41	42	43	44	45
46	47	48	49	50
51	52	53	54	55
56	57	58	59	60
61	62	63	64	65
66	67	68	69	70
71	72	73	74	75
76	77	78	79	80
81	82	83	84	85
86	87	88	89	90
91	92	93	94	95
96	97	98	99	100
101	102	103	104	105
106	107	108	109	110
111	112	113	114	115
116	117	118	119	120
121	122	123	124	125
126	127	128	129	130

You can check the interpretation of the question by first considering a specific case:

8	9
13	14

$8+9+13+14 = 44 = 4 \times 11$
$\rightarrow$ this case obeys the rule.

Generalising:
Let $n \in \mathbb{Z}^+$

n	$n+1$
$n+5$	$n+6$

Within any square, consider n to be the top-left value. Then complete the square in terms of n by using the relative position of the numbers.

Algebraically:

$$n+n+1+n+5+n+6 = 4n+12$$
$$= 4(n+3)$$

Showing that 4 can be removed as a factor demonstrates that the sum of the four numbers in the square is always a multiple of 4.

Therefore the sum of the numbers within a two-by-two square will always be a multiple of 4.

Proof by Exhaustion

Proof by exhaustion can be used when there are relatively few cases to check. It involves checking the rule with each case and, if they all work, the rule has been proved. Generally, questions will introduce limits on the numbers involved.

Example
Prove, by exhaustion, that the sum of any pair of different cube numbers less than 100 is not prime.

Cube numbers less than 100: 1, 8, 27, 64

For a number (n) to be prime, there has to be one pair of distinct factors (1 and n). If a number isn't prime, it will have at least one additional factor pair. Possible pairs:

$1 + 8 = 9$	As $9 = 3 \times 3$, it is not prime.
$1 + 27 = 28$	As $28 = 2 \times 14$, it is not prime.
$1 + 64 = 65$	As $65 = 5 \times 13$, it is not prime.
$8 + 27 = 35$	As $35 = 5 \times 7$, it is not prime.
$8 + 64 = 72$	As $72 = 2 \times 36$, it is not prime.
$27 + 64 = 91$	As $91 = 7 \times 13$, it is not prime.

Note: Any factor pair, other than 1 and n, will show that the number is not prime. As a step it needs to be shown: just saying '28 is not prime' misses out an important step. Equally it is worth stating the condition of primality/non-primality to complete the deductive steps. Otherwise it is a bit like giving someone directions and saying 'Once past the T-junction…' but not saying which way to turn at the junction.

Disproof by Counter-Example

Disproof is equally as important as proof. Instead of trying to prove that something is true, you could show that a theory cannot be true by using an example that shows it isn't true for all values.

Example
Cillian uses the equation $n^2 + 3n + 13$ to generate a sequence of numbers: 17, 23, 31, 41, … He says that all the numbers produced by this sequence will be prime numbers. Prove that Cillian is incorrect.

The sequence so far is all prime numbers. To find a number that isn't prime, it needs to have a common factor that is greater than 1 in each term of the expression. Given that the only factors of 13 are 1 and 13, a counter-example can be found using any multiple of 13, since this will give each term a common multiple of 13.

Let $n = 13$

$$n^2 + 3n + 13 = 13^2 + 3 \times 13 + 13 = 13(13 + 3 + 1)$$
$$= 13 \times 17$$

$\therefore n^2 + 3n + 13$, when $n = 13$, is not prime so Cillian is incorrect.

Any counter-example will do. For the previous example, any multiple of 13 would give at least a second pair of factors, e.g. $n = 26$ would give $n^2 + 3n + 13 = 767 = 13 \times 59$. Remember to complete a proof with a conclusion.

Sometimes disproof would form part of an amendment of the rule, for example by defining the possible set of values more accurately. The previous example shows that the equation will produce primes but not when n is a multiple of 13. This might still be useful to know and use in some contexts.

Don't forget
Sometimes a proof can seem to be so self-evident that 'because it is' feels like an appropriate response. In these cases, it is even more important to concentrate on the small steps and include all assumptions clearly.

Graphs and Diagrams

You can use graphs and diagrams as part of proofs. Using a calculator with a graphing function can help to simplify things, but it can also distract you. Don't expect the calculator to do the thinking for you and, if you are starting to use a trial and improvement approach, it could take some time. Detailed sketches are required to support working.

Links to Other Concepts
- Algebraic manipulation
- Number theory and sets
- Factorisation
- Indices
- Problem solving
- Polynomials
- Quadratics

SUMMARY

- You must be able to use and interpret set notation.
- Using appropriate and efficient notation to show working is key in helping to make work clear and concise.
- Types of proof required at AS-level:
  - Proof by deduction: a set of logical steps taken to get to the desired conclusion.
  - Proof by exhaustion: all possible values are tested against the rule and, if all work, then the rule is proved.
  - Disproof by counter-example: by finding one example that doesn't fit the rule, the rule is disproved.
- Showing all steps in a clear way supports the logical approach and helps others to interpret the work done. Use 'let', 'hence' and 'assuming' where relevant.
- Don't skip steps in a proof, even if they seem obvious.
- If a question says 'show that', or 'prove that', make sure the steps are detailed enough to count as a proof.

QUICK TEST

1. From the possible vales of n, where $n \in \{\sqrt{2}, 2, 3, \pi, 6, 7, 8, 8.\dot{3}, 9, 9.1, 12, \frac{62}{5}, 16\}$:
   a) List the integers.
   b) List the rational numbers.
   c) List the square numbers.
   d) List the prime numbers.
   e) List the irrational numbers.
2. List the factor pairs of 70.
3. Given that $x \in \mathbb{Z}^-$, state three possible values of x.
4. Write down, in algebraic terms, three consecutive integers.
5. Disprove the following statement: *The sum of two numbers is always greater than both of the numbers.*
6. a) Anand says that all primes are odd numbers. Prove that this is not true.
   b) Anand then says that all the primes between 10 and 50 are odd. Prove that he is correct.
7. Prove that the sum of any two consecutive square numbers is odd.

PRACTICE QUESTIONS

1. Prove that $(an+1)^2-(an-1)^2$, where $a \in \mathbb{Z}$, always gives an even number. **[4 marks]**
2. Charles has four coloured beads, as shown, to make a necklace. He ties a knot in one end and threads the purple bead first. If he selects the order randomly, prove that the probability that the blue and yellow beads will be next to each other is $\frac{2}{3}$. **[3 marks]**

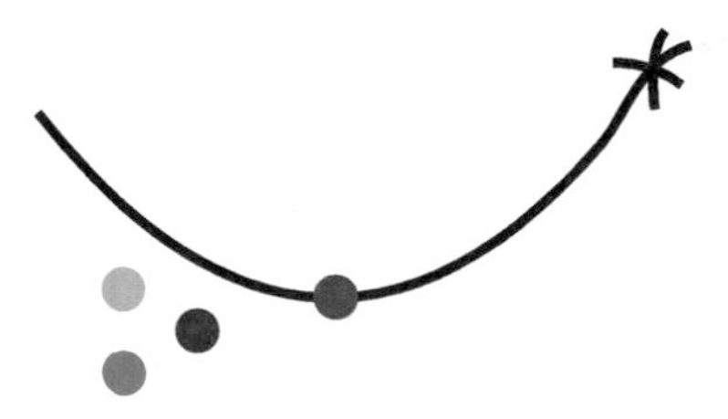

3. a) A number n is defined such that $n \in \{10, 14, 26, 70, 142, 206, 222, 254\}$.

   Billie says that $n = 2m$, where m is a prime number. Prove that Billie is incorrect. **[3 marks]**

   b) Billie says that a new number q is defined as $q \in \{2, 8, 18, 50, 60.5, 98\}$, and that q is always half of a square number.

   Prove whether or not Billie is incorrect. **[3 marks]**

Measures of Location and Spread

All measures of location and spread can be found using the statistical functions on your calculator. You will still need to be able to use the formulae, as sum data may be given to you rather than the raw data.

Class Boundaries

It is important that you consider the width of each group/class.

21–30
31–40
41–60

If this data is continuous (height, weight, etc.), the boundaries would be 20.5–30.5, 30.5–40.5, etc. Read the question carefully as it may tell you the data has been rounded to the nearest unit.

If the data is discrete (e.g. age) then the boundaries are as you see them, 21–30, etc.

$20 \leqslant h < 30$
$30 \leqslant h < 40$
$40 \leqslant h < 60$

Due to the inequality signs, the boundaries are 20–30, 30–40, etc. Notice the upper class boundary of the first class is the same as the lower class boundary of the second class. There is no value in-between to use as a limit.

Measures of Location

The **mean** is denoted by μ or $\bar{x}$. It is calculated using $\frac{\Sigma x}{n}$ where x is the data value and n is the number of values. For a frequency table, use $\frac{\Sigma fx}{\Sigma f}$. In a grouped frequency table, the x-value is the midpoint of the group.

Example
Find the mean of 3, 6, 5, 8, 5, 8, 8, 9.

$\Sigma x = 52$, $n = 8$, therefore $\mu = \frac{\Sigma x}{n} = \frac{52}{8} = 6.5$

Example
$\Sigma fx = 138$, $\Sigma f = 24$, therefore $\mu = \frac{\Sigma fx}{\Sigma f} = \frac{138}{24} = 5.75$

The **mode** is the most frequently occurring value.

Example
Find the mode of 3, 6, 5, 8, 5, 8, 8, 9.

The mode is 8.

Example

x	f
1	3
2	2
3	7

Find the mode.

Mode is 3 as it has the highest frequency.

Example

x	f
11–20	2
21–30	6
31–40	5

Find the modal class.

Modal group/class is 21–30 as it has the highest frequency.

The **median** is the middle value when the data is in size order. It is also known as the second quartile or Q_2. If there are n observations, find $\frac{n}{2}$. If $\frac{n}{2}$ is a whole number, find the midpoint of that term and the term above. If $\frac{n}{2}$ is not a whole number, round the number **up** and choose that term.

Example
Find the median of 3, 6, 5, 8, 5, 8, 8, 9.

First place the values in order: 3, 5, 5, 6, 8, 8, 8, 9

$\frac{n}{2} = 4$, a whole number, so $\frac{4^{\text{th}} + 5^{\text{th}}}{2}$ term, therefore the median is $\frac{6+8}{2} = 7$

Find the cumulative frequency (cf) to help with the position of the median.

Example

Find the median.

x	f	cf
1	3	3
2	4	7
3	8	15
4	2	17
5	4	21

$\frac{n}{2} = 11.5$, a decimal, so 12th term

The 8th to 15th terms are here, therefore the median is 3.

Example

Estimate the median height in this data.

Height (m)	f	cf
21–30	5	5
31–40	10	15
41–50	12	27
51–60	8	35

Again find the cumulative frequency to help with the position of the median, then interpolate $\frac{n}{2} = 17.5$. As this is an estimate, there is no need to round. The median is in the 41–50 group. Now use interpolation to estimate the median. Use the boundaries.

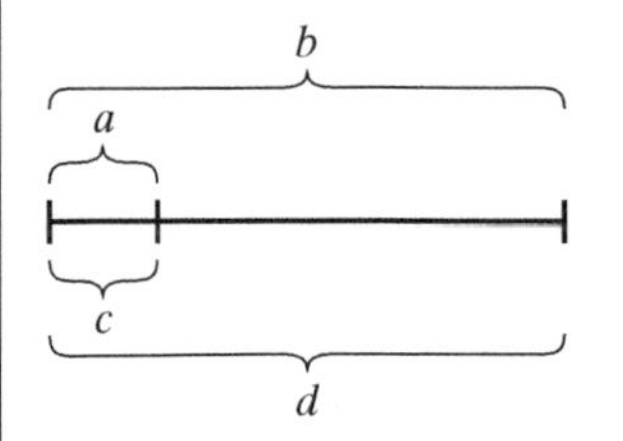

$$\frac{a}{b} = \frac{c}{d}$$

40.5 Q_2 50.5

15 17.5 27

$$\frac{Q_2 - 40.5}{50.5 - 40.5} = \frac{17.5 - 15}{27 - 15}$$

Rearrange and solve to give $Q_2 = 42.6$ m (3 s.f.)

Quartiles, **deciles** and **percentiles** all use the same methods as the median but the $\frac{n}{2}$ calculation needs to be changed for the position:

- For the lower quartile, Q_1, use $\frac{n}{4}$.
- For the upper quartile, Q_3, use $\frac{3n}{4}$.
- Deciles are tenths so for the third decile, D_3, use $\frac{3n}{10}$.
- Percentiles are hundredths so for the 51st percentile, P_{51}, use $\frac{51n}{100}$.

Measures of Spread

The **range** is the highest value – the lowest value.

The **interquartile range** is $Q_3 - Q_1$.

The **semi-interquartile range** is $\frac{Q_3 - Q_1}{2}$.

The **interpercentile range** will give two percentiles to find the difference between.

The **variance** is a calculation involving the squared distances of each value from the mean. It is often written as σ^2 but it can be written in many forms – check specific exam board formula books for the format given in the exam. The variance can also be found using a calculator.

$$\sigma^2 = \frac{\Sigma(x - \bar{x})^2}{n} = \frac{\Sigma x^2}{n} - \left(\frac{\Sigma x}{n}\right)^2 = \frac{S_{xx}}{n} \quad \text{where}$$

$$S_{xx} = \Sigma x^2 - \frac{(\Sigma x)^2}{n}$$

The **standard deviation** is the square root of the variance.

$$\sigma = \sqrt{\frac{\Sigma(x - \bar{x})^2}{n}} = \sqrt{\frac{\Sigma x^2}{n} - \left(\frac{\Sigma x}{n}\right)^2} = \sqrt{\frac{S_{xx}}{n}}$$

Example

Calculate the standard deviation of 3, 6, 5, 8, 5, 8, 8, 9.

$\Sigma x^2 = 368$, $\Sigma x = 52$, $n = 8$, therefore

$$\sigma = \sqrt{\frac{368}{8} - \left(\frac{52}{8}\right)^2} = 1.94 \text{ (3 s.f.)}$$

DAY 6

Coding

Data can be coded to make it easier to deal with. Numbers can be 'resized' to make calculations more manageable.

If the data was coded from original x-values to y by using the coding formula $y = \frac{x - a}{b}$:

- For the mean, calculate the mean of y ($\bar{y}$). The true mean will then be $\bar{x} = b\bar{y} + a$.
- For the standard deviation, calculate the standard deviation of y (σ_y). The true standard deviation will then be $\sigma_x = b\sigma_y$.

Addition and subtraction does not affect the spread of the data, therefore doesn't need uncoding.

Example

Data for 12 x-values was coded using the formula $y = \frac{x - 35}{20}$. It was found that $\sum y^2 = 1720$, $\sum y = 84$. Find the mean and the standard deviation of x.

First find the mean and the standard deviation of y:

$\bar{y} = \frac{84}{12} = 7$

Now uncode: $\bar{x} = 20\bar{y} + 35$
$= 20 \times 7 + 35$
$= 175$

$\sigma_y = \sqrt{\frac{1720}{12} - \left(\frac{84}{12}\right)^2} = 9.71253\ldots$ (use full answer)

Now uncode: $\sigma_x = 20\sigma_y$
$= 20 \times 9.71253\ldots$
$= 194.25$ (2 d.p.)

Interpreting Averages

For qualitative data (i.e. worded data, e.g. hair colour), only the mode can be used.

For quantitative data (i.e. numerical data), all other summary statistics can be used. If the data has extreme values (outliers), then the median and the interquartile range will be a more accurate representation of the data. However, without extreme values, the mean and the standard deviation should be used. As the mean and the standard deviation calculations use all of the data values, any outliers will make them misleading.

When interpreting data, observations should always be written in the context of the question and be backed up with the data.

Links to Other Concepts

- Representing data
- Statistical sampling

SUMMARY

- **The mean, μ or $\bar{x}$, is calculated by $\frac{\Sigma x}{n}$ or $\frac{\Sigma fx}{\Sigma f}$**
- **The mode/modal class has the highest frequency.**
- **The median is the middle value when the data is in size order.**
- **Quartiles split the data into quarters.**
- **Deciles split the data into tenths.**
- **Percentiles split the data into hundredths.**
- **Range = highest value – lowest value**
- **Interquartile range = $Q_3 - Q_1$**
- **Interpercentile range is the difference between two percentiles.**
- **Variance $\sigma^2 = \frac{\Sigma(x - \bar{x})^2}{n} = \frac{\Sigma x^2}{n} - \left(\frac{\Sigma x}{n}\right)^2 = \frac{S_{xx}}{n}$**
- **Standard deviation, σ, is the square root of the variance.**
- **When data is coded, the mean is affected by all four operations; the standard deviation is not affected by addition and subtraction.**

QUICK TEST

1. Find the mean and the variance for this data:

x	0.2	0.3	0.4	0.5	0.6	0.7
f	2	5	6	11	4	1

2. Ruth measured the lengths, to the nearest mm, of 100 leaves and recorded them here:

Length	25–29	30–34	35–39	40–49	50–69
Frequency	6	11	34	27	22

   a) Find the modal class.

   b) Estimate the mean leaf length.

3. Find the interquartile range for this data:

x	1	2	3	4	5	6
f	15	20	22	23	11	4

4. The heights of nursery children were measured, to the nearest cm, and recorded:

Height	60–69	70–79	80–89	90–109
Frequency	8	12	18	4

   Estimate the median height.

5. Find the 30th to the 80th interpercentile range for this data:

x	10	11	12	13	14	15
f	25	38	42	45	16	7

6. The mean age of 30 people at a birthday party was 28. After an hour, another five people joined the party; the mean age of those people was 34. Find the new average age of all the people at the party, to the nearest year.

7. Imran recorded test results (%) by gender and the results were as follows:

Female	7, 48, 49, 55, 58, 58, 60, 61, 62, 64, 64, 65
Male	18, 22, 56, 56, 57, 57, 57, 58, 60, 61, 63, 63, 63

   Compare the test results and comment on your findings.

8. Data was coded using the formula $p = \frac{q-12}{5}$.

   If $\bar{p} = 4$ and $\sigma_p = 3$, find the mean and the standard deviation of q.

PRACTICE QUESTIONS

1. The lengths of songs in the top 40 were timed to the nearest second in 1997 and again in 2017. The results are recorded in the table below.

Length	$120 \leqslant x < 150$	$150 \leqslant x < 180$	$180 \leqslant x < 190$	$190 \leqslant x < 200$	$200 \leqslant x < 240$	$240 \leqslant x < 270$	$270 \leqslant x < 350$
1997	8	15	7	5	2	3	0
2017	1	2	12	18	3	1	3

   a) Estimate the mean and the standard deviation of the length of songs in the top 40 for both years. **[4 marks]**

   b) Compare how song lengths have changed over time. **[2 marks]**

2. The number of litres, l, of milk produced one day by a herd of 60 Friesian cows was measured to the nearest litre. The following summary statistics were calculated:

   $\sum l = 1608, \ \sum l^2 = 53\,129$

   a) Calculate the mean and the standard deviation of the litres of milk produced daily. **[3 marks]**

   Unfortunately, one of the milking herd died. She had produced 32 litres on the day the data was collected.

   b) How will her death affect the mean milk production? **[2 marks]**

   c) Calculate the new $\sum l$ and $\sum l^2$. **[2 marks]**

   The number of litres of milk produced by 30 goats one day was also investigated. The number of litres, l, was also measured to the nearest litre and the data was coded using $y = 4(l + 3)$. The statistics found were:

   $\sum y = 1080, \ \sum y^2 = 41\,024$

   d) Find the mean and the standard deviation of the number of litres of milk produced by a goat in one day. **[5 marks]**

Probability

Probability is the likelihood of different events happening from an experiment. The probability of a particular event is found using $\frac{\text{number of required outcomes}}{\text{total number of outcomes}}$. Probabilities can be theoretical or estimated from an experiment. All probabilities lie between 0 and 1 and are usually written as fractions or decimals.

Some formulae are given in the exam formula booklet; make sure you check which are provided, as this varies depending on the exam board. The use of specific notation is not expected at AS-level but is a more concise way of expressing probabilities.

Experimental Probability

Example

A coin is flipped 100 times and lands on a head 63 times. On the other occasions, it lands on a tail. Find the probability of it landing on a tail.

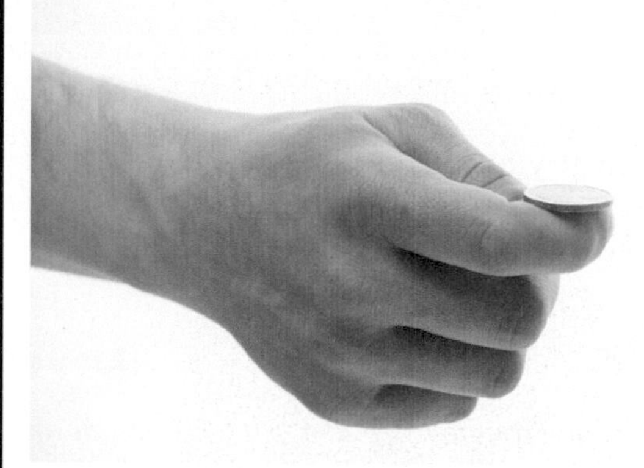

$P(\text{head}) = \frac{63}{100}$

The probability of not getting a head (i.e. getting a tail) is 1 – P(head).

$P(\text{tail}) = 1 - P(\text{head}) = 1 - \frac{63}{100} = \frac{37}{100}$

These events are called the **complement** of each other and can be written as $P(A') = 1 - P(A)$, where $P(A')$ is the probability of A not happening.

Finding Outcomes

It is useful to draw a sample space diagram to find all outcomes from an experiment.

Example

Two ordinary, fair tetrahedral dice are rolled and the product of the numbers obtained is the score. Find the probability of scoring greater than 10.

		Dice 1			
	×	1	2	3	4
Dice 2	1	1	2	3	4
	2	2	4	6	8
	3	3	6	9	12
	4	4	8	12	16

$P(>10) = \frac{3}{16}$

Tree Diagrams

The outcomes of successive events can be shown on a tree diagram with relevant probabilities on the branches. Probabilities are multiplied along the branches and then branch probabilities are added to give multiple combinations.

Example

Alia either walks to school or catches the bus. The probability that she walks is 0.6. If Alia walks, the probability of her being late is 0.2, but if she catches the bus the probability of her being late increases to 0.5. What is the probability that Alia is late on Monday?

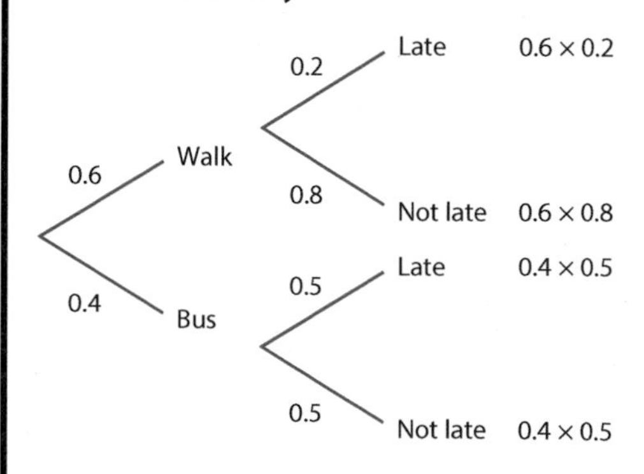

There are two options: Alia can walk and be late or go by bus and be late. Therefore:

$(0.6 \times 0.2) + (0.4 \times 0.5)$

$= 0.12 + 0.2 = 0.32$

Venn Diagrams

Events can be represented on a Venn diagram. Two or more events can be shown on a Venn diagram. Probabilities or frequencies can be written inside.

A **and** B ($A \cap B$) is also called the intersection and is represented by the shading.

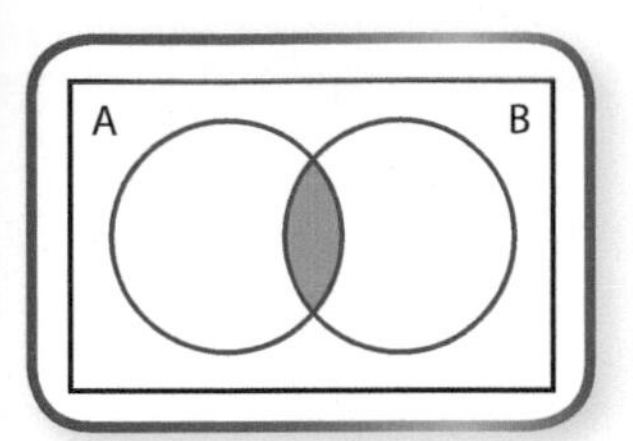

A **or** B ($A \cup B$) is also called the union and is represented by the shading.

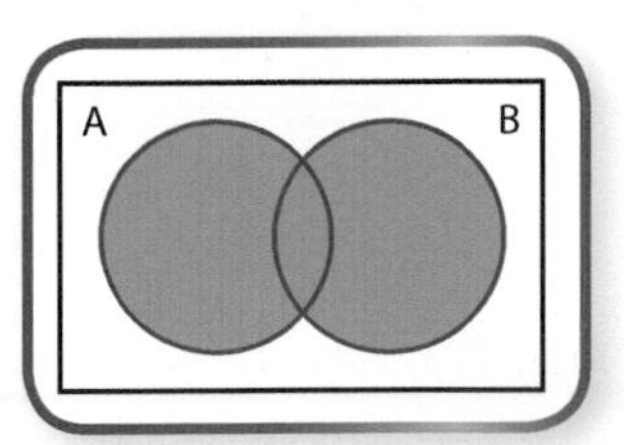

$P(A \cup B) = P(A) + P(B) - P(A \cap B)$

Mutually Exclusive Events

Events are mutually exclusive if they can't happen at the same time, i.e. they have no intersection. Therefore:

$P(A \cup B) = P(A) + P(B)$

A Venn diagram for two events A and B that are mutually exclusive looks like this:

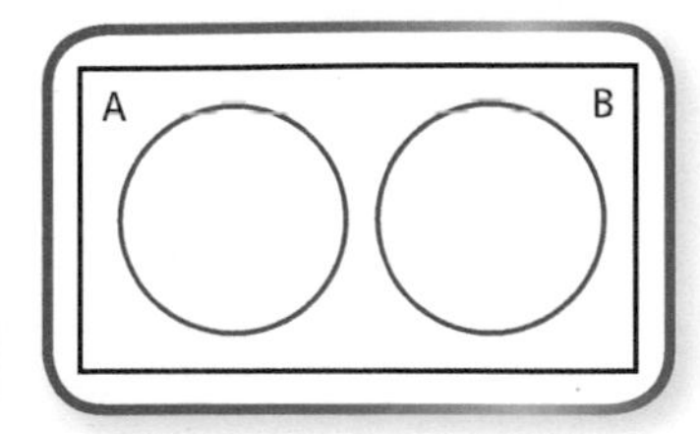

Independent Events

Events are independent if one event has no effect on the other, e.g. flipping a coin and rolling a dice. $P(A \cap B)$ is calculated using the multiplication rule for independent events, $P(A \cap B) = P(A) \times P(B)$. Events without replacement are not independent.

Links to Other Concepts

- Statistical distributions
- Hypothesis testing

SUMMARY

- **Probabilities add up to 1.**
- **$P(A') = 1 - P(A)$**
- **Sample space diagrams itemise each outcome.**
- **Venn diagrams represent events diagrammatically with frequencies or probabilities inside.**
- **Tree diagrams show outcomes of successive events.**
- **Mutually exclusive events have no intersection:**

  $P(A \cap B) = 0$
- **Independent events don't affect each other:**

  $P(A \cap B) = P(A) \times P(B)$

QUICK TEST

1. Events A and B are independent. If $P(A) = 0.3$ and $P(B) = 0.2$, calculate $P(A \cap B)$.
2. Events C and D are mutually exclusive. If $P(C) = 0.4$ and $P(C \cup D) = 0.7$, find $P(D)$.
3. 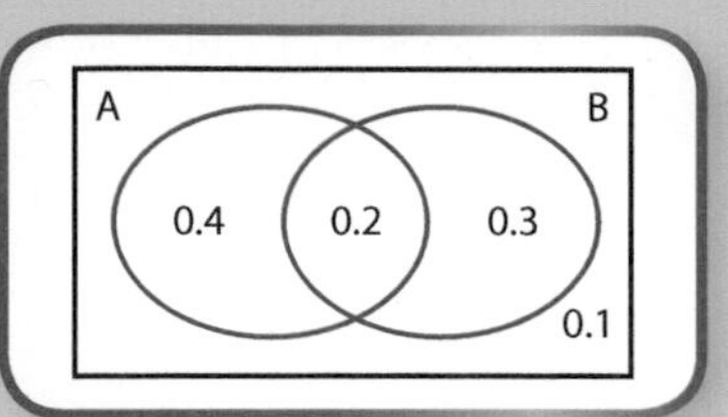

   Find:

   a) $P(A)$ **b)** $P(B')$ **c)** $P(A \cap B)$ **d)** $P(A \cup B)$

4. The tree diagram shows the probability of randomly taking a red or a blue sweet from a packet:

   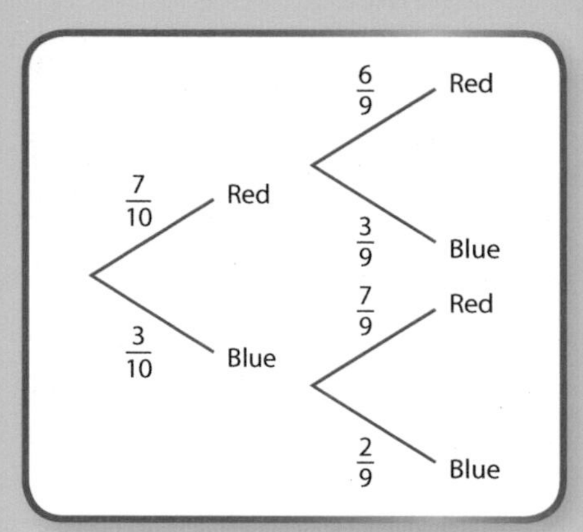

   One sweet is chosen at random and eaten, then a second one is chosen. What is the probability the sweets are different colours?

5. Two fair spinners are spun and the scores are added together. One spinner is numbered 1, 3, 5, 7 and the other 2, 4, 6, 8.

   Draw a sample space diagram to show all possible outcomes. What is the probability of scoring less than or equal to 7?

6. A dice was rolled and the results recorded in a table:

Number	1	2	3	4	5	6
Frequency	20	22	31	24	20	18

   What is the experimental probability of rolling a number 6? Is the dice fair?

7. Thirty students were surveyed:
   20 study Maths, 17 study Biology and 14 study both.
   Represent this information in a Venn diagram.

PRACTICE QUESTIONS

1. A particular type of biscuit is made in three different factories in England. Factories A, B and C produce 60%, 25% and 15% of the biscuits respectively. The proportion of broken biscuits is monitored and, on average, 5% of those from factory A are broken, 3% from factory B and 4% from factory C.

a) Represent the proportion of broken biscuits from the factories in a tree diagram. **[3 marks]**

b) If a biscuit is taken at random, find the probability that:

i) it is a broken biscuit from factory B **[2 marks]**

ii) it is not a broken biscuit. **[2 marks]**

2. Rebecca asked 70 people what genre of music they like to listen to from pop (P), rock (R) and classical (C). The results are given in the Venn diagram:

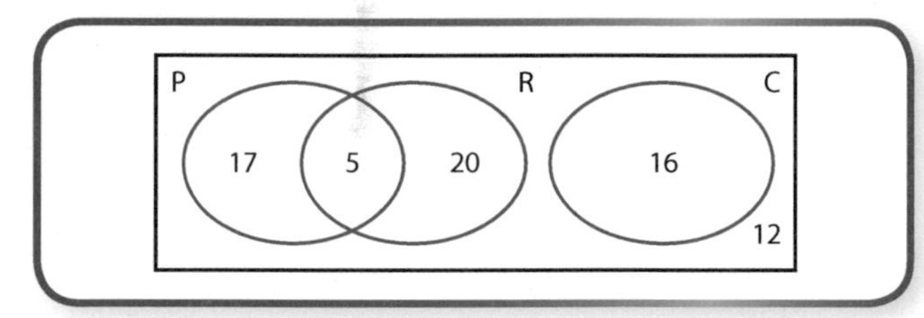

a) Write down two events that are mutually exclusive. **[1 mark]**

b) Find the probability that a person chosen at random likes rock and pop. **[1 mark]**

c) Are the events 'listening to pop music' and 'listening to rock music' statistically independent? Justify your answer. **[3 marks]**

3. An analysis of the gender of students was carried out in the sixth-form of Highway School. The results are shown in the two-way table:

	Year 12	Year 13
Girls	65	50
Boys	98	72

a) One student is chosen at random.
What is the probability that it is not a Year 12 boy? **[2 marks]**

b) Two students are chosen at random.
Find the probability that one is a girl and the other is a Year 13 boy. **[3 marks]**

Statistical Sampling

At AS-level it is important to be able to understand and use statistical terms and sampling methods, know the differences between them and critique them by understanding their advantages and disadvantages. Questions on this topic are likely to be related to the large data set used by the exam board.

Statistical Terms

A **population** in statistics is a complete set of items/events involved in a given investigation, e.g. all of the students in a sixth form.

A **sample** is a smaller set of these items chosen to represent the population. When 100% of the items are used, this is called a **census**. Using a sample to make inferences about a population is faster and cheaper than a census. However, depending on the sampling technique used, the sample may be biased.

Bias can be caused by who took the sample, where and when the sample was taken, how the data was collected and how big the sample was.

Sampling Techniques

Some of the most popular sampling techniques are as follows:

Simple Random Sampling

Every item in the population has an equal chance of being selected. Number each item in the population; this is your **sampling frame**. Then numbers are chosen at random using, for example, a random number generator on a calculator, or a raffle ticket approach (**lottery sampling**). Simple random sampling is an unbiased, fast and cheap method but is time consuming if a large sample is needed.

Systematic Sampling

Systematic sampling is another method for randomly choosing items where each item has an equal chance of being chosen. If a sample of 30 is needed from a population of 240, then choose every eighth item (240 ÷ 30) from a random starting point between 1 and 8. Systematic sampling is a fast and cheap method, even for large samples, but can be biased.

Stratified Sampling

The population is split into strata, such as Year 12 and Year 13 students. A sample is chosen in proportion to the size of the strata. If a sample of 30 is needed from a population of 240 students comprising 150 Year 12 students and 90 Year 13 students:

- Calculate $\frac{30}{240} \times 150 = 18.75$ for Year 12 and $\frac{30}{240} \times 90 = 11.25$ for Year 13.
- Choose 19 Year 12 and 11 Year 13 students using simple random sampling or systematic sampling. Always check your numbers add up to the sample size ($19 + 11 = 30$); if not, change your rounding.

Stratified sampling is representative of the population but can take a significant amount of time for larger strata.

Quota Sampling

In quota sampling, a sample is chosen specifically to represent different groups. Items are placed in turn into their specific group until the quota for that group is reached. This is a quick sampling method which represents all groups in the population but the non-random nature can cause bias in the sample.

Opportunity (Convenience) Sampling

Opportunity sampling is where the first items available are used as the sample, e.g. the first 30 students who walk into college on Monday. This method is cheap and easy to carry out but is likely to be biased and not representative of the population.

Links to Other Concepts

- Measures of location and spread
- Representing data
- Probability

QUICK TEST

1. Explain the difference between a census and a sample.
2. Abby asks the first 20 people going into a college library how often they read. What type of sampling technique has she used? Critique her choice of sampling method.
3. Mohammed wants to find the average salary of company employees. He takes a sample of 30 employees from a total of 360. Explain fully how he can use the method of systematic sampling to choose his sample.
4. Fishermen weigh each fish they catch in a competition and immediately pass the readings to the judge. The judge selects the first 10 salmon and the first 10 trout weights received to calculate an average weight of each breed. What type of sampling method is being used? Critique this choice of sampling method.
5. There are 160 boys and 100 girls in Year 12 of a college. Explain how you would take a stratified sample of 30 Year 12 students.

SUMMARY

- **A population is a whole set of items from a data set.**
- **A census looks at every member of a population.**
- **There are many sampling techniques, the most popular being:**
  - **simple random sampling**
  - **systematic sampling**
  - **stratified sampling**
  - **quota sampling**
  - **opportunity (convenience) sampling.**

Representing Data

Being able to interpret what a graph or chart shows is an important skill. All findings should be written in the context of the question.

Frequency Polygons

Frequency polygons are often used to compare multiple sets of continuous data. Frequencies are plotted against class midpoints and points joined with straight lines.

Example

Height (cm)	Boys	Girls
100–104	7	17
105–109	8	20
110–114	10	15
115–119	5	8
120–124	3	11

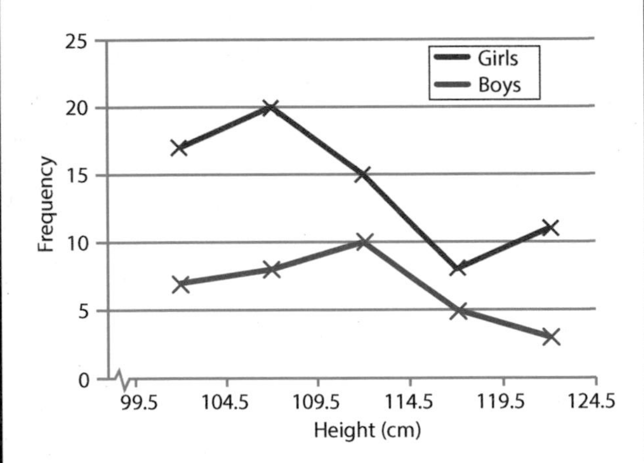

There are more girls than boys. The distribution of heights is similar, but the trend changes between 120 and 124 cm; the girls are on an upward trend but the boys are on a downward trend.

Cumulative Frequency Curves

A cumulative frequency curve generally represents one set of data but multiple curves can be drawn on one graph to compare multiple sets of data. The cumulative data is represented by an 'S'-shaped curve. Cumulative frequencies are plotted against upper class boundaries. Measures of location are found by drawing lines across and down on the graph.

Example

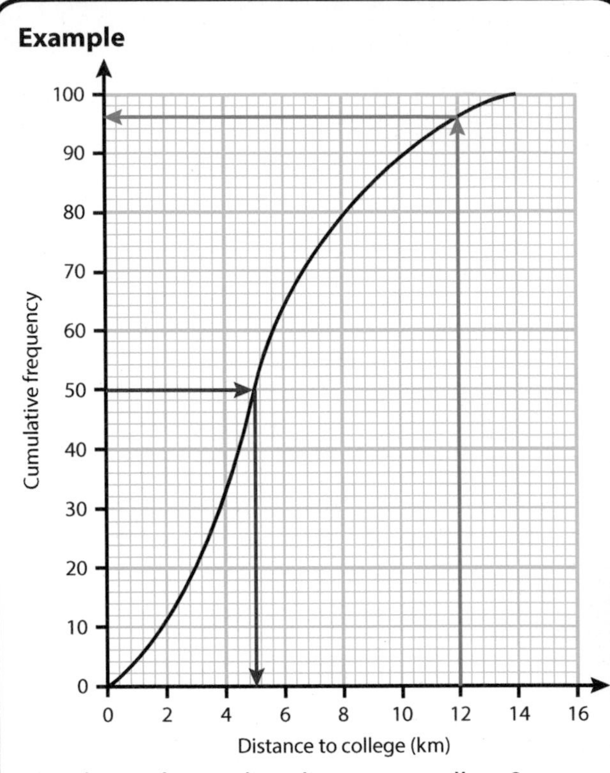

a) What is the median distance to college?

Cumulative frequency is 100 so the median is from 50 across and down (red lines).

Median = 5.1 km

b) How many students travelled further than 12 km to college?

Green lines drawn up from 12 km and across to 96.

100 – 96 = 4 students travelled more than 12 km.

Box Plots

Box plots show the spread of data between the lowest and highest values and the quartiles. More than one box plot can be drawn on each scale for comparisons. A box plot is shown. 25% of the data lies between each quartile.

Example

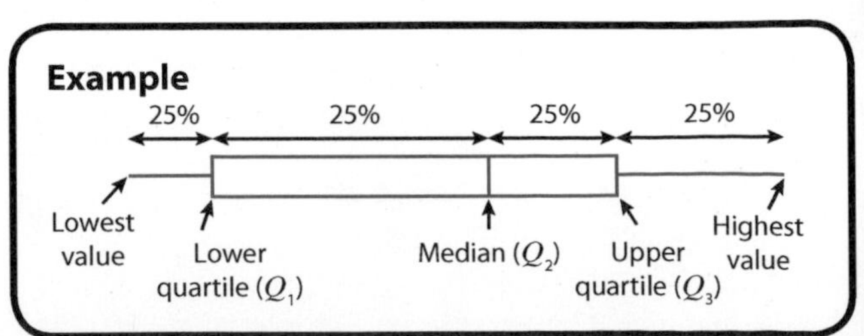

Outliers

Sometimes there are unusually high or low values in a set of data. There may be a good reason for this but sometimes they occur because an error was made when the data was recorded. Outliers are usually denoted using X or * on a box plot. There may be more than one outlier.

Method 1: Any value which is more than $k \times$ interquartile range (IQR) beyond the nearest quartile is classed as an outlier. k is often 1.5 but other values can be used. Any values above $Q_3 + 1.5 \times \text{IQR}$ and any below $Q_1 - 1.5 \times \text{IQR}$ are outliers.

Example
$Q_1 = 21, Q_3 = 33$
Is 5 an outlier? What about 55?

$1.5 \times \text{IQR} = 1.5 \times (33 - 21) = 18$

$Q_1 - 18 = 21 - 18 = 3$, so anything below 3 is an outlier.

$Q_3 + 18 = 33 + 18 = 51$, so anything above 51 is an outlier.

Therefore 5 is not an outlier but 55 is an outlier.

Method 2: Any value which is outside $\bar{x} \pm 2\sigma$ is classed as an outlier, or the question may state $\bar{x} \pm 3\sigma$. Approximately $\frac{2}{3}$ of the data lies within two standard deviations of the mean and 99% within three.

Example
10, 12, 12, 15, 16, 18, 20, 57
Are there any outliers?

$\bar{x} = 20, \sigma = 14.3$

$\bar{x} - 2\sigma = -8.6$, so no outliers below this.

$\bar{x} + 2\sigma = 48.6$, so 57 is an outlier as it is above 48.6.

Cleaning Data

An outlier can be a valid piece of data but sometimes we can confidently remove the outlier when we know it is an error. It would be misleading to leave this data in. We call these anomalies rather than outliers. Removing anomalies is called 'cleaning the data'.

Example
The speeds that cars pass school gates in a 30 mph zone are:
29, 31, 28, 31, 30, 30, 31, 30, 29, 30, 29, 28, 29, 28, 31, 32, 30, 28, 31, 86

Check for outliers:

$\bar{x} = 32.55, \sigma = 12.3$

$\bar{x} - 2\sigma = 7.95$, so no outliers below this.

$\bar{x} + 2\sigma = 57.15$, so 86 is an outlier and actually very unrealistic, therefore an anomaly. To clean the data, remove 86 before performing any statistical calculations.

Histograms

The frequency on a histogram is represented by the area of the bar. The vertical axis is the frequency density and is calculated using frequency density $= \frac{\text{frequency}}{\text{class width}}$.

Often questions give information about a section of the histogram and require other frequencies or the frequency density to be calculated.

Example
The histogram shows the distance travelled to college in km. 21 students travelled between 1 and 4.5 km. How many students were there altogether?

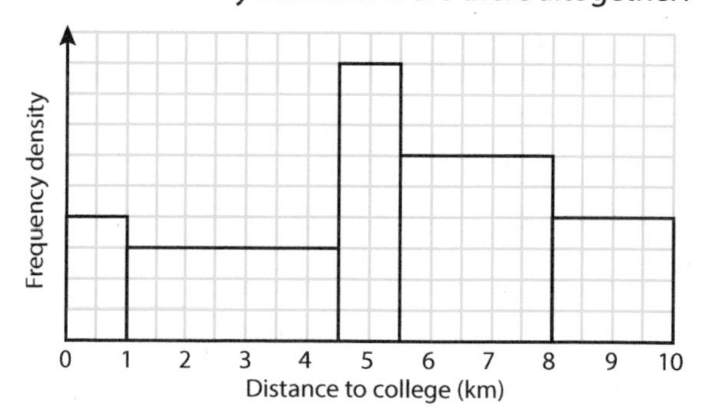

Area of 1–4.5 bar $= 3.5 \times h = 21$

$h = 6$, therefore each square on the frequency density axis is 2.

Total area
$= (8 \times 1) + (6 \times 3.5) + (18 \times 1) + (12 \times 2.5) + (8 \times 2) = 93$

DAY 7

There may be no diagram but information given about the histogram instead.

Example

In some continuous data, the frequency for the 10–29 group was 16. It was represented by a bar of horizontal width of 1 cm and a height of 4 cm.

The frequency for the 30–69 group was 28. What width and height should be used?

First, make a table with everything given in the question (or that can be calculated).

	Class width	Bar width	Frequency	Frequency density	Bar height
10–29	20	1 cm	16	16 ÷ 20 = 0.8	4 cm
30–69	40		28	28 ÷ 40 = 0.7	

The class width of 40 is double 20, therefore 1 cm × 2 = **2 cm**

The frequency density of 0.8 is multiplied by 5 therefore 0.7 × 5 = **3.5 cm**

Scatter Diagrams

Scatter diagrams show the relationship between two variables. The explanatory variable (independent) is on the horizontal axis and the response variable (dependent) is on the vertical axis. Relationships between the two variables are described as correlation but must be written in the context of the question.

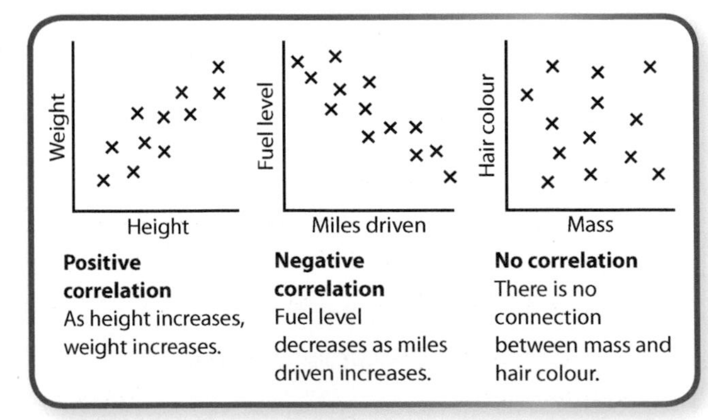

Positive correlation
As height increases, weight increases.

Negative correlation
Fuel level decreases as miles driven increases.

No correlation
There is no connection between mass and hair colour.

Correlation can be described further in terms of strong and weak positive and negative correlation. It is important to realise that correlation between two variables does not mean that one of them caused the other.

The regression line (line of best fit) may be given to help find approximate values for missing data items using the graph. It is written in the form $y = a + bx$. b can be interpreted as the change in the y unit for every one change in the x unit. Interpolation is finding estimates for values within the range of the data and extrapolation is estimating for values outside the range. Interpolation provides a reliable estimate but extrapolation is not reliable, therefore is not advised.

Example

A keen teenage runner times her best 100 m each year from the age (x) of 11. Her time (y) (in seconds) is modelled by the regression equation $y = 35 - 1.2x$. Interpret 1.2.

Think about interpreting a gradient in terms of 'y per x'. The negative means it is a decrease, so 'a decrease of 1.2 seconds in her 100 m time each year'.

SUMMARY

- **Outliers are extreme values of data. Calculate using quartiles:**

  $Q_1 - 1.5 \times \text{IQR}$ $\quad$ $Q_3 + 1.5 \times \text{IQR}$

  or mean and standard deviation:

  $\bar{x} \pm 2\sigma$
- **Cleaning data is the removal of anomalies.**
- **For histograms, frequency density** $= \frac{\text{frequency}}{\text{class width}}$
- **Correlation describes the relationship between two variables.**
- **The regression line is the line of best fit,** $y = a + bx$.

Links to Other Concepts
- Measures of location
- Measures of spread
- Statistical sampling

QUICK TEST

1. For the data below, determine whether there are any outliers.

   18, 12, 15, 38, 23, 17, 25, 21, 23, 19, 20, 20, 18

2. At the start of a week-long adult training camp, participants ran 100 m, then ran it again at the end of the week. Their times were:

1st run	10.9	11.3	11	10.7	10.8	10.5	11.2	11.1
2nd run	10.8	10.9	10.9	10.7	10.9	10.4	11.1	11

The regression line is $y = 3.41 - 0.85x$.

a) Interpret the coefficient of x.

b) Angel is a primary school teacher who wants to use this model in her PE lessons. A child with a first run time of 21 seconds wants to know by how much she will improve in a week. Is this a suitable model to use?

3. Describe the relationship between the age of a car and its mileage.

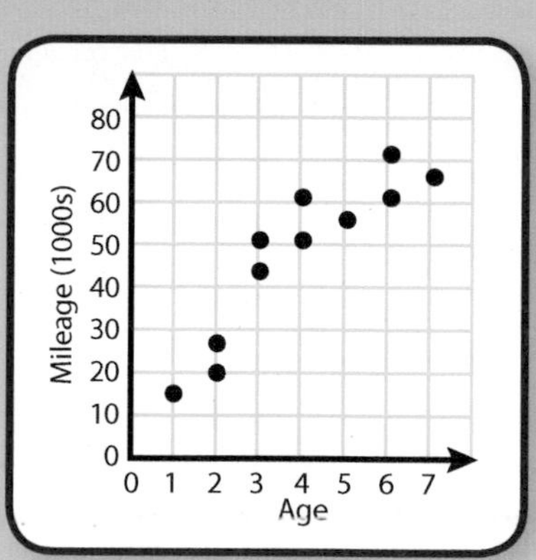

4. Find the median and the interquartile range.

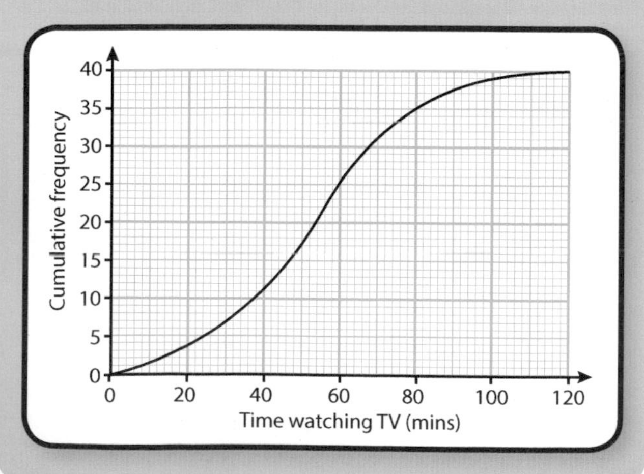

PRACTICE QUESTIONS

1. Maths test results are shown below.

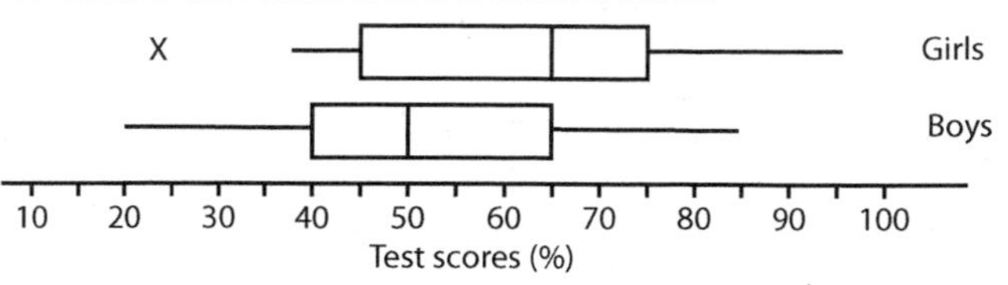

a) Sandeep thinks that 75% of the girls scored above 65%. Is he correct? Explain your answer fully. **[1 mark]**

b) What does the cross represent? **[1 mark]**

c) If 75% of the boys passed the test, what was the pass mark? Explain how you know. **[2 marks]**

d) Compare the boys' and girls' results. **[2 marks]**

2. Jenny's A-level class all planted some sunflower seeds and waited for them to grow. The heights of the sunflowers were then measured in cm and the results put into a histogram:

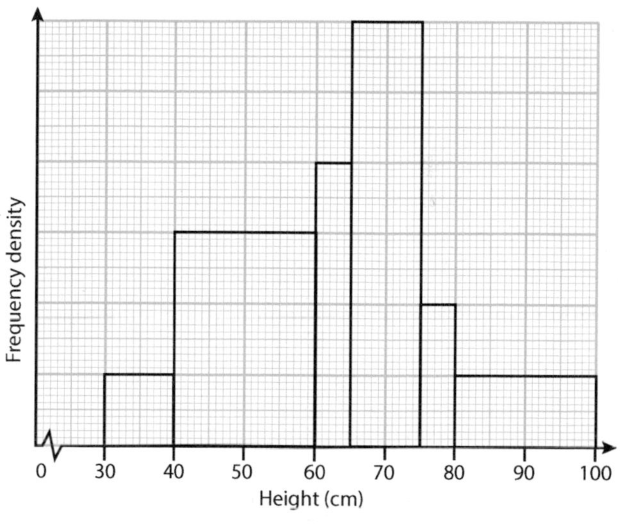

a) 35 of the sunflowers were between 65 and 80 cm tall. How many of the sunflowers grew between 40 and 60 cm tall? **[3 marks]**

b) How many sunflowers were planted altogether? **[2 marks]**

Statistical Distributions

Discrete probability distributions can be used to model everyday situations. This topic covers simple discrete probability distributions and specifically the binomial distribution.

Discrete Probability Distributions

A discrete probability distribution can be written in a variety of forms, e.g. as a function of x, as a table or as a diagram. It shows the outcomes from a probability experiment involving a discrete random variable. The discrete random variable is described using a capital letter and each individual value it takes using a lowercase letter.

A function of x: $P(X=x)=\frac{x}{6}$ for $x=1,\ 2,\ 3$

A table:

x	1	2	3
$P(X=x)$	$\frac{1}{6}$	$\frac{1}{3}$	$\frac{1}{2}$

A diagram:

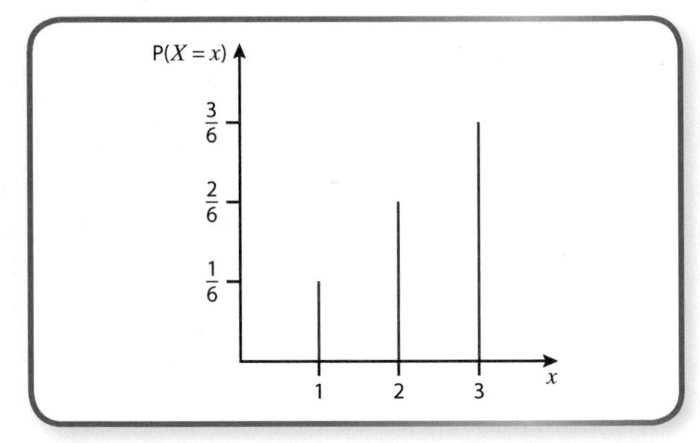

Probabilities add up to 1. This is written as $\sum P(X=x)=1$. This fact can be used to find missing probabilities.

Example
Find the value of c.

x	3	6	9
$P(X=x)$	0.1	c	0.3

$c=1-(0.1+0.3)$

$c=0.6$

Example
Find the value of y: $P(X=x)=\begin{cases}\frac{x}{8} & \text{for } x=1,3\\ y & \text{for } x=5\end{cases}$

$y=1-\left(\frac{1}{8}+\frac{3}{8}\right)$

$y=\frac{4}{8}=\frac{1}{2}$

The above example can be converted from its function of x into table form. It is often easier to use in this format:

Example

x	1	3	5
$P(X=x)$	$\frac{1}{8}$	$\frac{3}{8}$	$y=\frac{1}{2}$

Questions involving probabilities can then be answered.

Example
Find $P(X \geqslant 3)$ for this distribution.

$P(X=3)+P(X=5)=\frac{7}{8}$

Exam questions can involve situations where prior probability knowledge is required to form the distribution.

> **Example**
> A dice is rolled until it lands on a 6 or until it has been rolled three times. Write down the probability distribution for X, 'the number of times the dice is rolled'.
>
> So, the dice could be rolled one, two or three times. Probabilities need to be calculated for each of these:
>
> $P(x=1)=\frac{1}{6}$
>
> If the dice is only rolled once, then a 6 must have been rolled.
>
> $P(X=2)=\frac{5}{6}\times\frac{1}{6}=\frac{5}{36}$
>
> If the dice is only rolled twice, it must have landed on a number other than 6 on the first roll and on a 6 on the second roll.
>
> $P(X=3)=\left(\frac{5}{6}\times\frac{5}{6}\times\frac{1}{6}\right)+\left(\frac{5}{6}\times\frac{5}{6}\times\frac{5}{6}\right)=\frac{25}{36}$
>
> If the dice is rolled three times, there are two options, hence the addition: 'not 6', 'not 6' and a 6 on the third roll **or** 'not 6', 'not 6' and 'not 6'.
>
> This is then written as a probability distribution in a table:

x	1	2	3
$P(X=x)$	$\frac{1}{6}$	$\frac{5}{36}$	$\frac{25}{36}$

Discrete Uniform Distribution

Where a discrete probability distribution has equal probabilities for all of the outcomes, it is called a discrete uniform distribution.

> **Example**
> Find $P(X\leqslant 5)$.

x	1	2	3	4	5	6	7	8	9
$P(X=x)$	$\frac{1}{9}$	$\frac{1}{9}$	$\frac{1}{9}$	$\frac{1}{9}$	$\frac{1}{9}$	$\frac{1}{9}$	$\frac{1}{9}$	$\frac{1}{9}$	$\frac{1}{9}$

> $P(X\leqslant 5)=5\times\frac{1}{9}=\frac{5}{9}$

There is no expectation to draw the probability distribution (table) if it is not given. Multiply the number of options by the probability as all probabilities are equal.

The Binomial Distribution

The binomial distribution is a discrete distribution which models the outcomes of an experiment that repeats a number of times. There are specific conditions that define whether a situation does follow the binomial distribution. These are:

- A fixed number of trials, n.
- Two possible outcomes: success and failure.
- A constant probability of success, p.
- The trials are independent.

A binomial distribution is written as $X\sim B(n, p)$, hence one with 20 trials and a probability of success of $p=0.2$ is written as $X\sim B(20, 0.2)$.

Calculating Probabilities

Probabilities are calculated using the formula:

$P(X=r)=\binom{n}{r}p^r(1-p)^{n-r}$, where $\binom{n}{r}={}^nC_r=\frac{n!}{r!(n-r)!}$

This is not provided on all exam board formula sheets as probabilities can be found on A-level specific calculators.

> **Example**
> If $X\sim B\left(10,\frac{1}{3}\right)$, find $P(X=2)$.
>
> $P(X=2)=\binom{10}{2}\left(\frac{1}{3}\right)^2\left(\frac{2}{3}\right)^8=\frac{1280}{6561}=0.195$ (3 s.f.)

Cumulative tables – if using these, note that the probabilities given in the tables are all $P(X\leqslant r)$.

Calculators – there are two functions: a probability, PD, that is $P(X=r)$, which could have been used for the example above; and a cumulative probability, CD, that is $P(X\leqslant r)$. For the following examples, choose the 'variable' option.

If a question asks for $P(X \geq r)$, calculate $1 - P(X \leq r-1)$.

If a question asks for $P(X > r)$, calculate $1 - P(X \leq r)$.

Answers can be rounded to 4 decimal places from calculators to match cumulative tables.

Example
If $X \sim B(20, 0.2)$, find $P(X \geq 5)$.

$P(X \geq 5) = 1 - P(X \leq 4) = 1 - 0.6296 = 0.3704$

Example
35% of chocolates in a tin have soft centres. 30 chocolates are chosen at random. Find the probability that more than half of the chocolates have soft centres.

Half of 30 is 15 so find $P(X > 15)$.

This is the same as $1 - P(X \leq 15) = 1 - 0.9699 = 0.0301$

Working Backwards

Sometimes a probability is given and the r value needs to be found. This can be more straightforward on tables, if they are provided, as the probabilities in the correct column can be scanned to find those either side of the required probability. To use the calculator function, work with the cumulative probabilities but use the 'list' rather than the 'variable' option. Values for r need to be input into the list; there is an element of trial and error involved.

Example
If $M \sim B(20, 0.3)$, find m such that $P(M \geq m) \leq 0.05$.

$P(M \geq m) = 1 - P(M \leq m-1)$

Check different values of m:

$P(M \geq 9) = 1 - P(M \leq 8) = 0.1133$

$P(M \geq 10) = 1 - P(M \leq 9) = 0.048$, therefore $m = 10$

Links to Other Concepts
- Probability
- Hypothesis testing

SUMMARY

- **A discrete random variable can only take certain values.**
- **Probabilities add up to 1:** $\sum P(X = x) = 1$
- **A discrete uniform distribution has equal probabilities.**
- **A binomial distribution has the properties:**
  - **A fixed number of trials, n.**
  - **Two possible outcomes: success and failure.**
  - **A constant probability of success, p.**
  - **Independent trials.**
- **Binomial probabilities are calculated using the calculator distribution function or** $P(X = r) = \binom{n}{r} p^r (1-p)^{n-r}$.

QUICK TEST

1. Are the following discrete random variables?
   a) The distance, X, a child travels to school, in metres
   b) The number of heads thrown, X, when a coin is tossed 10 times
2. The discrete random variable Y has probability function:

   $f(y)=cy$ for $y=1, 2, 3$

   Find the constant c.
3. The discrete random variable R has probability distribution:

R	3	5	7	9
$P(R=r)$	$\frac{1}{12}$	$\frac{1}{4}$	k	$\frac{1}{4}$

   Find $P(R>6)$.
4. A binomial distribution has 10 trials and a probability of success of 0.2. Find the probability of having four successes.
5. $X \sim B(30, 0.15)$. Find:
   a) $P(X \leqslant 7)$ b) $P(X<4)$
   c) $P(X>10)$ d) $P(X \geqslant 8)$
6. Find y such that $P(Y \leqslant y)=0.2969$ when $Y \sim B(15, 0.3)$.
7. $M \sim B(25, 0.4)$. Find $P(3 \leqslant M<8)$.
8. A discrete uniform distribution for the discrete random variable X can take any value between 1 and 20 inclusive. Find $P(X<10)$.
9. The discrete random variable N has the probability distribution:

$$f(n)=\begin{cases} k(n^2+2) & n=0, 2, 4, 6 \\ kn & n=8, 10, 12 \\ 0 & \text{otherwise} \end{cases}$$

   Find k.
10. Amani flips a fair coin 20 times. Find the probability that she gets at least eight heads.

PRACTICE QUESTIONS

1. A door-to-door double glazing salesman has a record of convincing 70% of the homeowners in Redmine to have a free quote.
   a) One afternoon the salesman knocks at 50 homes in Redmine. What is the probability that he convinces more than 36 homeowners to have a free quote? **[2 marks]**
   b) The following day the salesman knocks at 50 homes in Downland. Will the probability of him convincing more than 36 homeowners to have a free quote be the same as the day before? Explain your answer. **[2 marks]**
2. The random variable Y has the probability function:

$$P(Y=y)=\begin{cases} \frac{y}{4} & y=1, 2 \\ ky & y=3, 4 \end{cases}$$

   a) Find the value of k. **[2 marks]**
   b) Draw a probability distribution table for Y. **[2 marks]**
   c) Find $P(2 \leqslant Y<4)$. **[1 mark]**
3. A driving school advertises a 95% first-time pass rate for students. The Driving Test Standards Agency samples 30 people who have taken their driving test for the first time in the last year using that driving school.
   a) Explain how the sample could be taken. **[1 mark]**
   b) Find the probability that the sample contains fewer than three people who fail if the pass rate advertised by the driving school is correct. **[2 marks]**

   The Driving Test Standards Agency will investigate the driving school if more than f people fail in the sample.

   c) Given the agency has to investigate 1.56% of the time on average, find the value of f. **[2 marks]**

Hypothesis Testing

At AS-level, hypothesis tests are carried out using the binomial distribution. A hypothesis is a theory that is made about the distribution parameter. In the case of the binomial this is p, the probability of success.

Terminology

H_0 is the **null hypothesis**. This is what we think is true. A decision for the test is based on accepting or rejecting H_0. We always assume that H_0 is correct and we only reject it if there is enough evidence to.

H_1 is the **alternative hypothesis**. This claims something other than the null hypothesis is true.

H_0 is always written as $p = \ldots$

One-tailed tests only look at one end of the data, so for H_1, $p < \ldots$ or $p > \ldots$

Two-tailed tests look at both ends of the data, so for H_1, $p \neq \ldots$. The significance level is halved for two-tailed tests.

The **significance level** is the probability that you base your decision on. Common ones are 1%, 5% and 10%. It is the probability of incorrectly rejecting H_0.

The **actual significance level** can be different to the significance level the test was conducted at. It is the probability that a value lies within the critical region.

The **test statistic** is a number you are given within the problem. The probability of this value is found to determine whether to reject or accept H_0.

The **critical region** is the region in which the test statistic would be rejected.

The **critical value** is the first value which lies within the critical region.

Writing Hypotheses

Always start a question by defining H_0 and H_1.

Example

a) A dice is rolled 40 times and it lands on a 6 four times. Test at the 5% significance level whether the dice is biased against 6s.

H_0: $p = \frac{1}{6}$

(the theoretical probability of rolling a 6)

H_1: $p < \frac{1}{6}$

If the dice is biased against 6s, the probability of rolling a 6 will be less than $\frac{1}{6}$; so this is a one-tailed test.

b) A dice is rolled 40 times and it lands on a 6 four times. Test at the 5% significance level whether the dice is biased.

H_0: $p = \frac{1}{6}$, H_1: $p \neq \frac{1}{6}$

Notice the change of wording; this time the test is two-tailed as the question doesn't specify in which way the dice is biased.

Conducting a Test in General

All required probabilities can be calculated using the statistical function on suitable calculators, as in the previous topic. This is recommended as, depending on the exam board, the printed tables may not be in the formula booklet.

Conducting a Test – Using a Test Statistic

The alternative hypothesis is mirrored when deciding what probability to calculate.

For H_1, $p < \ldots$ find $P(X \leqslant \text{test statistic})$, for $p > \ldots$ find $P(X \geqslant \text{test statistic})$ and for $p \neq \ldots$ both probabilities can be calculated, although it is often obvious which one to use from the position of the test statistic within the distribution compared to np (the mean). If this probability is less than the significance level, then H_0 is rejected and a conclusion is written. Where there is a context in the question, the conclusion must be written accordingly.

Example

From the distribution $X \sim B(30, 0.2)$, with a test statistic of 11, test at the 5% significance level the hypotheses H_0: $p = 0.2$ and H_1: $p > 0.2$.

$P(X \geqslant 11) = 0.0256$ (4 d.p.)

This is less than the 5% level so H_0 is rejected.

As $0.0256 < 0.05$, reject H_0 and conclude it is likely that p is greater than 0.2.

Example

From the distribution $X \sim B(20, 0.45)$, with a test statistic of 5, test at the 10% significance level the hypotheses H_0: $p = 0.45$ and H_1: $p \neq 0.45$.

$P(X \leqslant 5) = 0.0553$ (4 d.p.)

As this is a two-tailed test, the 10% significance level is halved; the probability is more than half of the 10% significance level so H_0 is accepted.

As $0.0553 > 0.05$, accept H_0 and conclude there is no reason to suggest p is not 0.45.

Conducting a Test – Critical Regions

Another way to conduct a test is to find the critical region for the test. Some questions specifically ask for the critical region to be found. If a test statistic is given in the question, either method can be used. Calculators (or tables if given) are used to find the first value which has a probability less than the significance level (again halved for a two-tailed test). This is the critical value and the critical region is formed from this, again following the inequality on the alternative hypothesis. There are two critical regions for a two-tailed test.

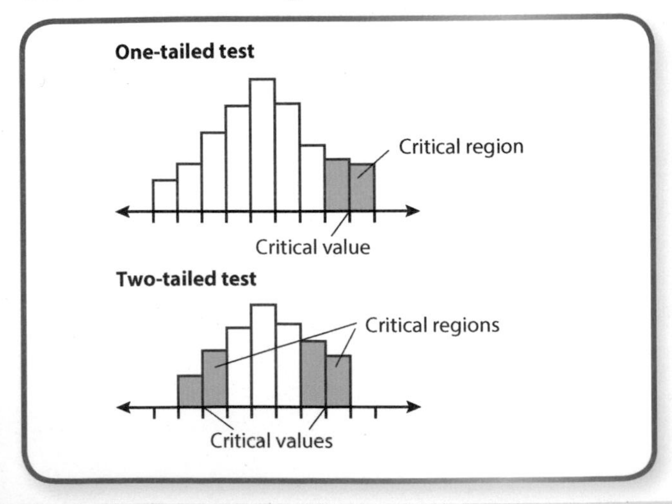

Example

From the distribution $X \sim B(50, 0.15)$, find the critical region for a test at the 10% significance level for the hypotheses H_0: $p = 0.15$ and H_1: $p < 0.15$.

$P(X \leqslant 3) = 0.0460$

$P(X \leqslant 4) = 0.1121$

Look at $\leqslant$ probabilities as H_1 is '<'. Always show values either side of the significance level.

The critical region is $X \leqslant 3$.

This is the first value which gives a probability less than the significance level; 3 would be the critical value.

Example

From the distribution $X \sim B(25, 0.4)$, find the critical region for a test at the 5% significance level for the hypotheses H_0: $p = 0.4$ and H_1: $p \neq 0.4$.

$P(X \leqslant 4) = 0.0095$

$P(X \leqslant 5) = 0.0294$

$P(X \geqslant 15) = 1 - P(X \leqslant 14) = 0.0344$

$P(X \geqslant 16) = 1 - P(X \leqslant 15) = 0.0132$

The critical region is $X \leqslant 4$ and $X \geqslant 16$. These are the first values which give probabilities less than half of the significance level at each end of the distribution.

The actual significance level of the test in the previous example is $0.0095 + 0.0132 = 0.0227$, found by adding the probabilities for the critical regions. This is also referred to as the probability of incorrectly rejecting H_0.

If tables are available, find values for the previous example as shown:

$p=$	0.05	0.10	0.15	0.20	0.25	0.30	0.35	0.40	0.45	0.50
$n=25, x=$ 0	0.2774	0.0718	0.0172	0.0038	0.0008	0.0001	0.0000	0.0000	0.0000	0.0000
1	0.6424	0.2712	0.0931	0.0274	0.0070	0.0016	0.0003	0.0001	0.0000	0.0000
2	0.8729	0.5371	0.2537	0.0982	0.0321	0.0090	0.0021	0.0004	0.0001	0.0000
3	0.9659	0.7636	0.4711	0.2340	0.0962	0.0332	0.0097	0.0024	0.0005	0.0001
4	0.9928	0.9020	0.6821	0.4207	0.2137	0.0905	0.0320	0.0095	0.0023	0.0005
5	0.9988	0.9666	0.8385	0.6167	0.3783	0.1935	0.0826	0.0294	0.0086	0.0020
6	0.9998	0.9905	0.9305	0.7800	0.5611	0.3407	0.1734	0.0736	0.0258	0.0073
7	1.0000	0.9977	0.9745	0.8909	0.7265	0.5118	0.3061	0.1536	0.0639	0.0216
8	1.0000	0.9995	0.9920	0.9532	0.8506	0.6769	0.4668	0.2735	0.1340	0.0539
9	1.0000	0.9999	0.9979	0.9827	0.9287	0.8106	0.6303	0.4246	0.2424	0.1148
10	1.0000	1.0000	0.9995	0.9944	0.9703	0.9022	0.7712	0.5858	0.3843	0.2122
11	1.0000	1.0000	0.9999	0.9985	0.9893	0.9558	0.8746	0.7323	0.5426	0.3450
12	1.0000	1.0000	1.0000	0.9996	0.9966	0.9825	0.9396	0.8462	0.6937	0.5000
13	1.0000	1.0000	1.0000	0.9999	0.9991	0.9940	0.9745	0.9222	0.8173	0.6550
14	1.0000	1.0000	1.0000	1.0000	0.9998	0.9982	0.9907	0.9656	0.9040	0.7878
15	1.0000	1.0000	1.0000	1.0000	1.0000	0.9995	0.9971	0.9868	0.9560	0.8852
16	1.0000	1.0000	1.0000	1.0000	1.0000	0.9999	0.9992	0.9957	0.9826	0.9461
17	1.0000	1.0000	1.0000	1.0000	1.0000	1.0000	0.9998	0.9988	0.9942	0.9784
18	1.0000	1.0000	1.0000	1.0000	1.0000	1.0000	1.0000	0.9997	0.9984	0.9927
19	1.0000	1.0000	1.0000	1.0000	1.0000	1.0000	1.0000	0.9999	0.9996	0.9980
20	1.0000	1.0000	1.0000	1.0000	1.0000	1.0000	1.0000	1.0000	0.9999	0.9995
21	1.0000	1.0000	1.0000	1.0000	1.0000	1.0000	1.0000	1.0000	1.0000	0.9999
22	1.0000	1.0000	1.0000	1.0000	1.0000	1.0000	1.0000	1.0000	1.0000	1.0000

Questions can ask for the probabilities for the critical region to be 'as close as possible to' the significance level. The previous example would be reworded as follows.

Example

From the distribution $X \sim B(25, 0.4)$, for the hypotheses H_0: $p = 0.4$ and H_1: $p \neq 0.4$, find the critical region for a test at the 5% significance level. The probability in each tail should be as close as possible to 2.5%.

$P(X \leqslant 4) = 0.0095$

$P(X \leqslant 5) = 0.0294$

$P(X \geqslant 15) = 1 - P(X \leqslant 14) = 0.0344$

$P(X \geqslant 16) = 1 - P(X \leqslant 15) = 0.0132$

The critical region is now $X \leqslant 5$ and $X \geqslant 15$ (as 0.0294 is closer to 0.025 than 0.0095 is, and 0.0344 is closer to 0.025 than 0.0132 is).

Note: Only use this method when it is clearly stated in the question.

SUMMARY

- **The null hypothesis, H_0, is what we think is true, always written as $p = \ldots$**
- **The alternative hypothesis, H_1, claims something other than the null hypothesis is true.**
- **One-tailed tests are H_1, $p < \ldots$ or $p > \ldots$**
- **Two-tailed tests are H_1, $p \neq \ldots$**
- **The significance level is the probability of incorrectly rejecting H_0.**
- **The actual significance is the probability that a value lies within the critical region.**
- **The test statistic is a number given within the problem.**
- **The critical region is the region where the test statistic would be rejected.**
- **The critical value is the first value lying within the critical region.**

Links to Other Concepts

- Probability
- The binomial distribution

QUICK TEST

1. A coin is flipped 20 times and lands five times on heads. Tracy wants to test at the 1% significance level whether the coin is biased. Write her hypotheses.
2. If $X \sim B(15, p)$, find the critical region for a test at the 5% significance level for the hypotheses H_0: $p = 0.5$ and H_1: $p < 0.5$.
3. From the distribution $X \sim B(12, p)$, with a test statistic of 1, test at the 10% significance level the hypotheses H_0: $p = 0.3$ and H_1: $p \neq 0.3$.
4. If $X \sim B(25, p)$, find the critical region for a test at the 1% significance level for the hypotheses H_0: $p = 0.45$ and H_1: $p \neq 0.45$.
5. $X \sim B(10, p)$. Test at the 2% significance level the hypotheses H_0: $p = 0.35$ and H_1: $p > 0.35$. Use a test statistic of 8.
6. Explain what is meant by a hypothesis test.
7. Find the probability of incorrectly rejecting the null hypothesis when testing hypotheses H_0: $p = 0.2$ and H_1: $p < 0.2$ for $Y \sim B(50, p)$, at the 1% significance level.
8. Find the actual significance level of the test where the probability in each tail is as close as possible to 5% for $X \sim B(40, p)$, H_0: $p = 0.35$ and H_1: $p \neq 0.35$.

PRACTICE QUESTIONS

1. A new drug produced by pharmaceutical company Hastozena is tested as a treatment for migraines. Hastozena claims the drug reduces pain within 20 minutes for 30% of the population. A rival drug company thinks it works for less than 30% of the population.

   Forty people took part in the trial and six people agreed their pain was reduced within 20 minutes.

   Test Hastozena's claim at the 5% significance level. **[5 marks]**

2. The probability that Oakfield hockey team score their first goal of a game in the second half is 0.8. Their best goalscorer was injured before the last season and in the next 50 games Oakfield scored their first goal in the second half of 35 games.

   a) Test, at the 10% significance level, whether the proportion of first goals in the second half of each game has changed. **[5 marks]**

   b) State one assumption you made before conducting the test. **[1 mark]**

3. Plant genetics determine that $\frac{1}{4}$ of a certain species of rose bushes have red flowers. A random sample of 20 bushes was planted. It was found that nine bushes have red flowers. By considering the critical region, test at the 1% significance level whether the proportion of bushes with red flowers has increased. **[5 marks]**

4. Lisa claims that one-fifth of the chocolates produced by an old machine in a factory are mis-shaped. To test this claim, the number of mis-shaped chocolates in a random sample of 100 is recorded.

   a) Why might a binomial distribution be a suitable model for the number of mis-shaped chocolates in the sample? Give two reasons. **[2 marks]**

   b) Using a 5% significance level, find the critical region for a two-tailed test of the hypothesis that the probability of a chocolate being mis-shaped is $\frac{1}{5}$. The probability in each tail should be as close as possible to 2.5%. **[3 marks]**

   c) Find the actual significance level of this test. **[2 marks]**

Answers

Day 1

Indices and Surds

QUICK TEST (page 7)

1. $\frac{1}{3}$ **2.** $2x^3 - 2x^2$ **3.** $\frac{\sqrt{x}}{2}$ **4.** $\frac{\sqrt{6}}{9}$

5. $\frac{x\sqrt{2}+8}{x^2-32}$ **6.** C **7.** $2^{-\frac{3}{2}}$ **8.** $(3x^2 - 12)\sqrt{3} + 3x$

PRACTICE QUESTIONS (page 7)

1. Find expression for length AC: $\sqrt{12^2+8^2} = \sqrt{208}$ **[1]**
Divide DF by AC: $\frac{14}{\text{AC}} = \frac{14}{\sqrt{208}}$ **[1]**
Answer simplified to surd involving $\sqrt{52}$: $\frac{14}{2\sqrt{52}}$ **[1]**
Rationalised denominator $\frac{7\sqrt{52}}{52}$ **[1]**

2. For equating coefficients **[1]**
From x^2 coefficient, $c = 7$ **[1]**
$a\sqrt{8} = 10 \times 2^{\frac{1}{2}} \rightarrow 2a\sqrt{2} = 10\sqrt{2}$ **[1]**
$a = 5$ **[1]**
$\frac{16}{\sqrt{b}} = 2^{\frac{1}{2}} \times 2 \rightarrow \frac{16}{\sqrt{b}} = 2\sqrt{2} \rightarrow 16 = 2\sqrt{2} \times \sqrt{b}$, or other correct first step to solution **[1]**
$b = 32$ **[1]**

3. a) For using a multiplier of $\frac{a-\sqrt{5}}{a-\sqrt{5}}$ **[1]**
For a final answer equivalent to $\frac{4a\sqrt{5}-20}{a^2-5}$ **[1]**

b) For $27^{\frac{5}{2}} = (3\sqrt{3})^5 = 3^5 \times \sqrt{3}^5 = 3^7\sqrt{3}$ **[2]**
For $3^{-\frac{3}{2}} = \frac{1}{\sqrt{3}^3} = \frac{1}{3\sqrt{3}} = \frac{1}{9}\sqrt{3}$ **[2]**
For $3^{\frac{1}{2}} = 1 \times \sqrt{3}$ **[1]**
$(3^7 - 3^{-2} + 1)\sqrt{3}$, $a = 7$, $b = -2$ **[1]**

Polynomials

QUICK TEST (page 13)

1. $21x^2 - 13x - 18$ **2.** $3b^2 + 5b - a^2$ **3.** $12x^3 - 10x^2 + 11x - 3$

4. 0 **5.** A–R, B–Q, C–P, D–S

6. a) Two real and distinct roots
b) Two roots in the same place or one real repeated root
c) Two real and distinct roots
d) No real roots

7. $2.25 = \frac{9}{4}$ and $-2.25 = -\frac{9}{4}$ **8.** $p = 2, q = 10$

PRACTICE QUESTIONS (page 13)

1. a) $f\left(\frac{1}{2}\right) = 12\left(\frac{1}{2}\right)^3 + 5\left(\frac{1}{2}\right) - 4 = 0$, $\therefore$ $(2x - 1)$ is a factor **[1]**

b) **[1]**

$$\begin{array}{r} 6x^2 + 3x + 4 \quad [1] \\ 2x-1\overline{)\,12x^3 + 0x^2 + 5x - 4} \\ \underline{12x^3 - 6x^2} \qquad\qquad \\ 6x^2 + 5x \qquad \\ \underline{6x^2 - 3x} \qquad \\ 8x - 4 \\ \underline{8x - 4} \\ 0 \quad [1] \end{array}$$

c) Attempt to use discriminant $b^2 - 4ac$ **[1]**
$3^2 - 4 \times 6 \times 4 < 0$ $\therefore$ no real roots **[1]**

2. a) $x^2 + \frac{4}{3}x + \frac{1}{6} = 0$ **[1]**
$\left(x + \frac{2}{3}\right)^2 - \left(\frac{2}{3}\right)^2 + \frac{1}{6} = 0$ **[1]**
$\left(x + \frac{2}{3}\right)^2 = \frac{5}{18}$
$x + \frac{2}{3} = \pm\sqrt{\frac{5}{18}}$ **[1]**
$x = -\frac{2}{3} \pm \sqrt{\frac{5}{18}} = -\frac{2}{3} \pm \frac{\sqrt{5}}{3\sqrt{2}} = -\frac{4}{6} \pm \frac{\sqrt{5\times2}}{6}$ **[1]**
$= \frac{-4\pm\sqrt{10}}{6}$ **[1]**

b) Rearrange to form $6a^{1.2} + 8a^{0.6} + 1 = 0$ **[1]**
From part **a)**, $a^{0.6} = \frac{-4+\sqrt{10}}{6}$ or $a^{0.6} = \frac{-4-\sqrt{10}}{6}$ **[1]**
For a method of finding a. (Alternate method to taking the 0.6-root is to raise both sides to power $\left(\frac{5}{3}\right)$.) **[1]**
$a^{0.6} = \frac{-4+\sqrt{10}}{6} \rightarrow a = \sqrt[0.6]{\frac{-4+\sqrt{10}}{6}} \rightarrow a = -0.0376$ **[1]**
$a^{0.6} = \frac{-4-\sqrt{10}}{6} \rightarrow a = \sqrt[0.6]{\frac{-4-\sqrt{10}}{6}} \rightarrow a = -1.34$ **[1]**

3. a) For substitution of values into equation and rearranging to form quadratic $-1.8 = 10t - 5t^2$
$\rightarrow 5t^2 - 10t - 1.8 = 0$ or $-5t^2 + 10t + 1.8 = 0$ **[1]**
Solutions from any quadratic solving method and acknowledgement of result $-0.1661...$ **[1]**
For final solution of $t = 2.16619...$ seconds rounded to any suitable degree **[1]**

b) For identifying the different values: $s = 1.8$, $u = 10$, $a = -10$, $t = ?$ **[1]**
For the quadratic formed $5t^2 - 10t + 1.8 = 0$ **[1]**
For pair of solutions to equation $t = 0.2$ and $t = 1.8$ **[1]**
For finding time above $t = 1.8 - 0.2 = 1.6$ seconds **[1]**

Simultaneous Equations and Inequalities

QUICK TEST (page 19)

1. a) $t < -2$ **b)** $t \in (t : t < -2)$ or $t \in (-\infty, -2)$

2. $a = 3, b = -2$

3. a)

−2 −1 0 1 2 3 4 5 6 7 8 9

b) $x \leqslant 2, 5 \leqslant x < 8$

4. $a = 4$ **5.** −2, −1, 0, 1

6. (−1, −5) and (6, 2) **7.** $q \geqslant 5, q \leqslant \frac{3}{2}$

PRACTICE QUESTIONS (page 19)

1. a)

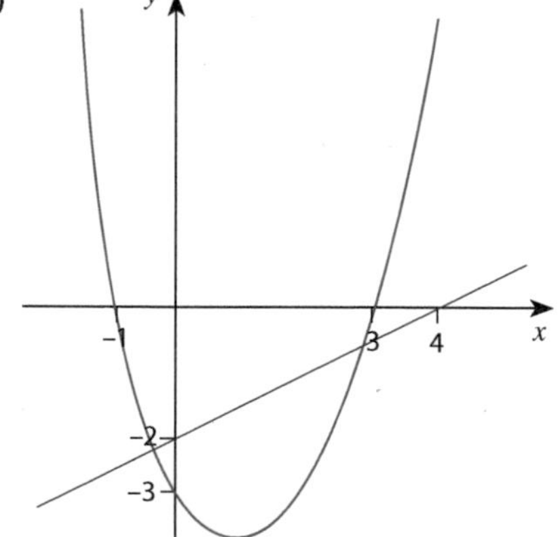

For straight line with positive gradient passing through y-axis below the origin **[1]**
For curve in correct orientation with two distinct x-intercepts **[1]**
For attempt to find x-intercepts for quadratic either by factorisation or other quadratic method $(x + 1)(x - 3)$ **[1]**
For all axis-intercepts labelled correctly **[1]**
For two clear intercepts of the two lines identified, both intercepts are below the x-axis so have a negative y-coordinate **[1]**

b) For method to solve as simultaneous equations (substitution or elimination both valid) **[1]**
$x^2 - 2x - 3 = \frac{1}{2}x - 2 \rightarrow 2x^2 - 4x - 6 = x - 4 \rightarrow 2x^2 - 5x - 2 = 0$ **[1]**

2.

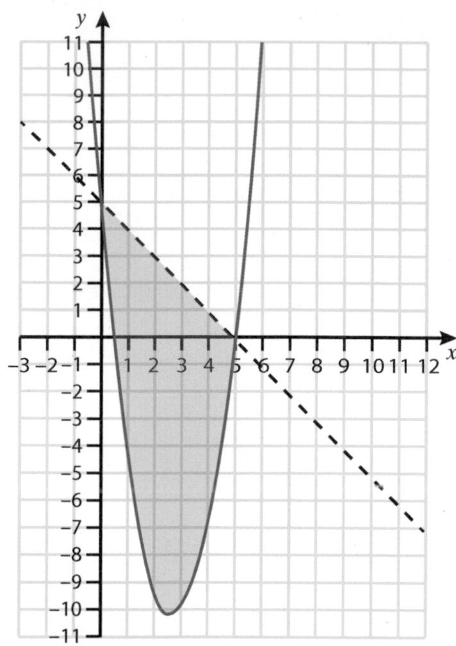

For correct line, dotted with shading only below **[1]**
For quadratic curve with correct x-intercepts and y-intercept **[1]**
For minimum point on the graph **[1]**
For correct region shaded on the graph **[1]**

3. Let x = number of pencils and y = number of pens
Tom: $2x + 5y = 3.45$, Dilys: $4x + 2y = 3.30$
For setting up the equations **[1]**
Eliminate (or substitute) to get $8y = 6.90 - 3.30 \rightarrow 8y = 3.60$ (or other valid first step) **[1]**
For solution for one of the unknowns, $y = 0.45$ **[1]**
Correct substitution into either equation
$2x + 5 \times 0.45 = 3.45 \rightarrow x = 0.60 \rightarrow x + y = £1.05$ **[1]**

4. Possible algebraic solution:
$ax - b < 0$ gives the result $x < 7$, only inequality in given solution that is < (rather than ⩽) **[1]**
$ax < b \rightarrow x < \frac{b}{a} \rightarrow 7 = \frac{b}{a}$ **[1]**
Positive bound from quadratic is from $\left(x - \frac{b}{a+5}\right) = 0$, as a and b are both positive
$x = \frac{b}{a+5} \rightarrow \frac{b}{a+5} = 2$ **[1]**
Rearrange one equation to isolate unknown $b = 7a$ **[1]**
Substitute into second equation $\frac{7a}{a+5} = 2$ **[1]**
Rearrange and solve for one unknown
$7a = 2a + 10 \rightarrow 5a = 10 \rightarrow a = 2$ **[1]**
Find second unknown $b = 7 \times 2 = 14$ **[1]**

Possible diagrammatic solution:

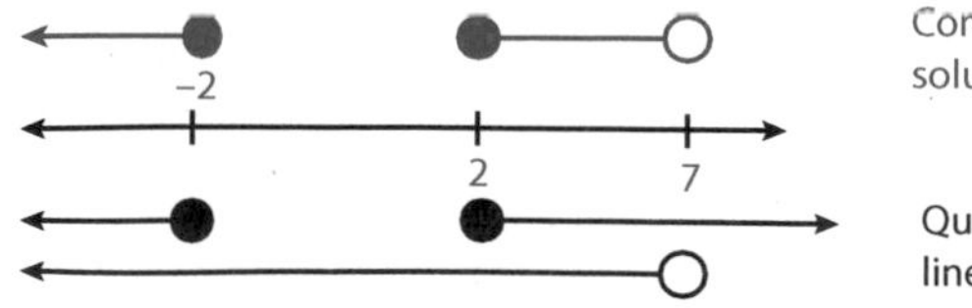

Combined solution

Quadratic linear **[2]**

$\left(x - \frac{b}{a+5}\right)$ $ax - b < 0$

$(x + 2)$ $\frac{b}{a+5}$ $\frac{b}{a}$

−2 2 7

$\frac{b}{a+5} = 2$ and $\frac{b}{a} = 7$ **[1]**
Rearrange one equation to isolate unknown $b = 7a$ **[1]**
Substitute into second equation $\frac{7a}{a+5} = 2$ **[1]**
Rearrange and solve for one unknown
$7a = 2a + 10 \rightarrow 5a = 10 \rightarrow a = 2$ **[1]**
$b = 7a = 14$ **[1]**

Graphs and Transformations

QUICK TEST (page 23)

1.

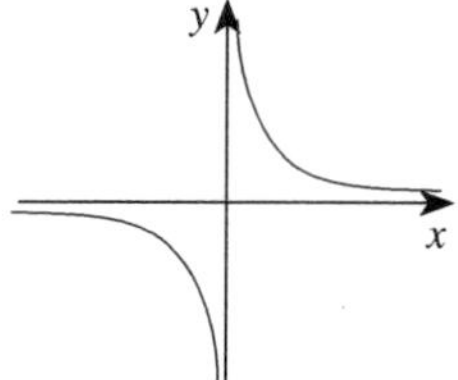

2. $x = -1$

3.

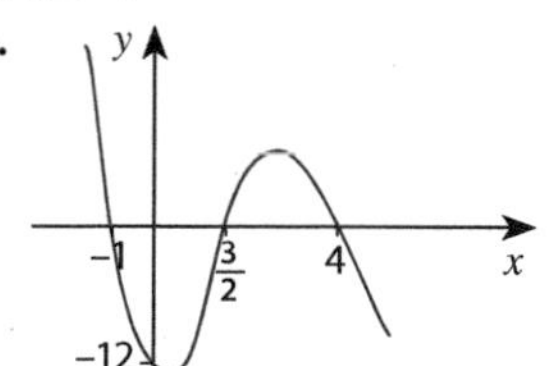

4. a) and b) $f(x + 1)$ so translation of $\binom{-1}{0}$ and $f(x) + 2$, which is a translation of $\binom{0}{2}$.

5.

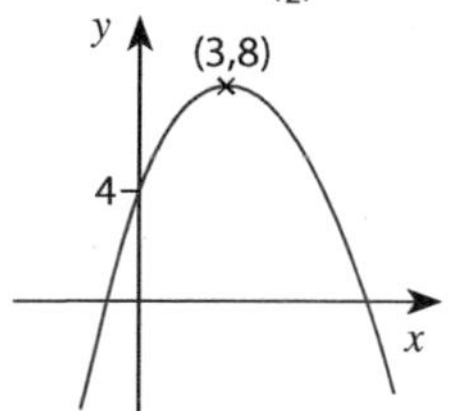

PRACTICE QUESTIONS (page 23)

1. a)

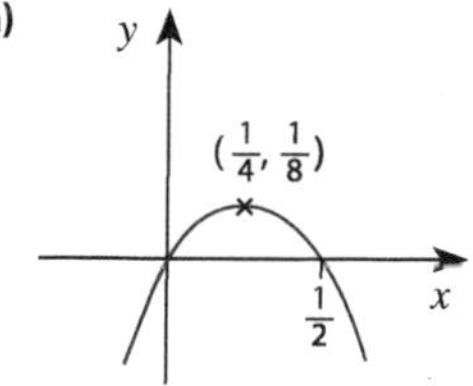

For shape of curve **[1]** For x-intercepts **[1]**

b) For finding the original maximum point, can be evident from part **a)**. Maximum y-value $= \frac{1}{8} = 0.125$ **[1]**
$\binom{0}{-0.125}$, or equivalent description. $y = f(x) - \frac{1}{8}$ **[1]**

2. a) $f(-1) = (-1)^3 - 3 \times (-1)^2 + 4 = -1 - 3 + 4 = 0$ **[1]**

$\therefore (x + 1)$ is a factor **[1]**

b) By algebraic division:

$$
\begin{array}{r|l}
 & x^2 - 4x + 4 \quad \leftarrow \textbf{[1]} \\
x + 1 & x^3 - 3x^2 + 0x + 4 \\
 & \underline{x^3 + x^2} \\
 & -4x^2 + 0x \quad \leftarrow \textbf{[1]} \\
 & \underline{-4x^2 - 4x} \\
 & 4x + 4 \quad \leftarrow \textbf{[1]} \\
 & \underline{4x + 4} \\
 & 0
\end{array}
$$

By equating coefficients:

$x^3 - 3x^2 + 4 = (x + 1)(ax^2 + bx + c)$

$x^3 - 3x^2 + 4 = ax^3 + (a + b)x^2 + (b + c)x + c$ **[1]**

$a = 1, c = 4$ **[1]**

$b = -4$ **[1]**

Final mark for full factorisation:

$y = (x + 1)(x^2 - 4x + 4) = (x + 1)(x - 2)^2$ **[1]**

c)

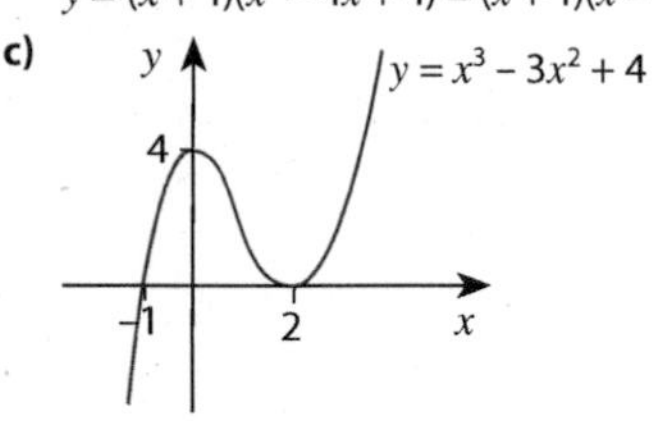

For a correct shape and direction **[1]** For correct axis-intercepts **[2]**

d) $f(x - 2) = (x - 2)^3 - 3(x - 2)^2 + 4$

For correct transformation applied (–2 within bracket) **[1]**

$= (x^2 - 4x + 4)(x - 2) - 3(x^2 - 4x + 4) + 4$

$= x^3 - 6x^2 + 12x - 8 - 3x^2 + 12x - 12 + 4$

For expanding the brackets **[1]**

$= x^3 - 9x^2 + 24x - 16$

For simplification and correct answer **[1]**

e)

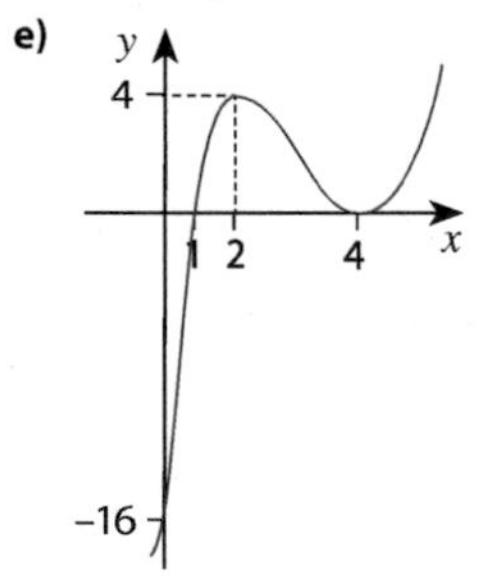

For correct shape with curve just touching x-axis at the minimum **[1]**

For correct axis-intercepts **[1]**

3. a) (0, 12) **[1]**

b) i)

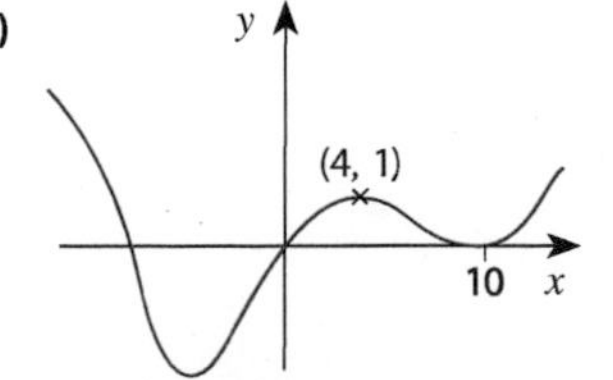

For $a = -10$ **[1]**

For a correct sketch passing through the origin **[1]**

For labelling the point $x = 10$ and showing the graph just touching the x-axis at this point **[1]**

ii) Translation of $\binom{5}{0}$ and $\binom{7}{0}$

A full description or expressed as $f(x - 5)$ and $f(x - 7)$ are also acceptable forms **[2]**

Day 2

Coordinate Geometry

QUICK TEST (page 27)

1. Gradient = 5
2. Length = 17 and gradient = $\frac{-8}{15}$
3. $(x+3)^2 + (y+1)^2 = 49$
4. Gradient is perpendicular = $\frac{-1}{2}$
5. $x + 2y = 0$ or $y = \frac{-1}{2}x$
6. 

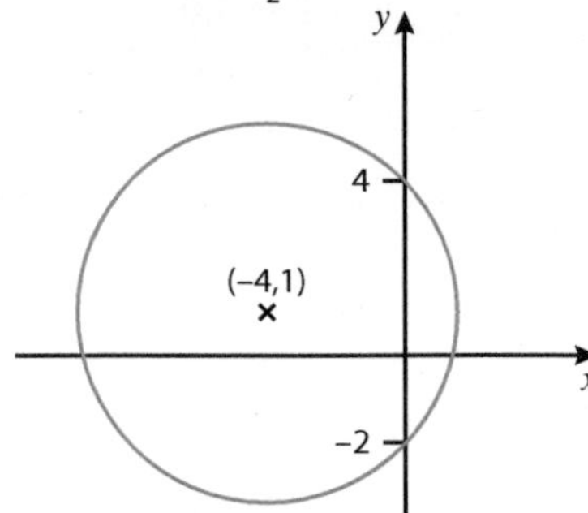

7. (1, 1) and (–1, –3)

PRACTICE QUESTIONS (page 27)

1. a)

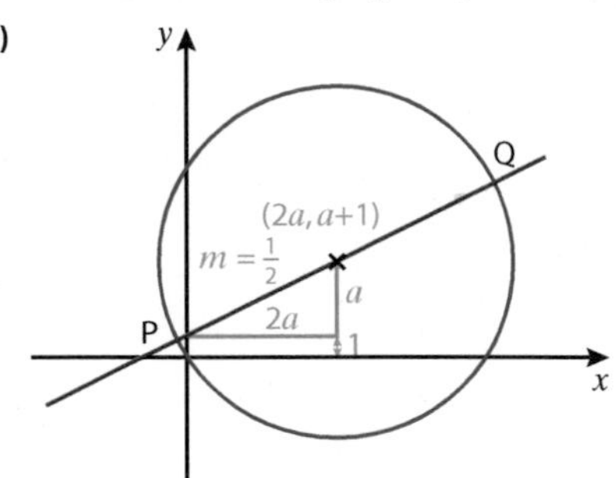

For use of the gradient of the line to get a ratio of sides of the right-angled triangle of 2 to 1 **[1]**

For demonstrating where the +1 comes from for the y-coordinate **[1]**

b)

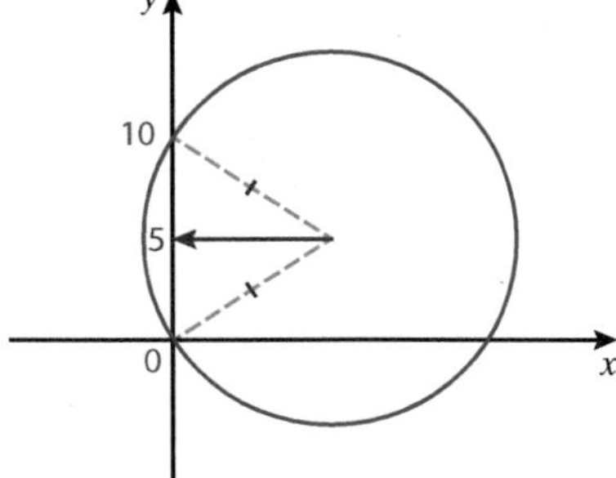

Use of isosceles triangle, or similar, to show the y-coordinate of the centre of the circle is at 5 **[1]**

$a + 1 = 5 \rightarrow a = 4$

Centre of the circle is at $(2a, a + 1) = (8, 5)$ **[1]**

c) Equation of circle given centre: $(x - 8)^2 + (y - 5)^2 = r^2$ **[1]**

Use of coordinates of centre of circle to find the radius

$r^2 = 8^2 + 5^2 = 89$ **[1]**

For fully correct equation (can be expanded but doesn't need to be) $(x - 8)^2 + (y - 5)^2 = 89$ **[1]**

(If expanded, final mark for: $x^2 + y^2 - 16x - 10y = 0$)

d) Length PQ is the diameter of the circle

$r = \sqrt{89}$

Diameter $= 2\sqrt{89}$ **[1]**

2. a) Set up equations for finding midpoint $\left(\frac{2k+-2}{2}, \frac{3+(k-3)}{2}\right)$ or $m = \frac{2k+-2}{2}$ and $m = \frac{3+(k-3)}{2}$ **[1]**

$\rightarrow \quad \frac{2k-2}{2} = \frac{k}{2}$ **[1]**

$2k - 2 = k$

$k = 2$ **[1]**

b) $2\sqrt{13}$ **[1]**

c) Gradient of AB $= \frac{2}{3}$ **[1]**

Gradient of BC $= \frac{-3}{2}$ **[1]**

Substitution into $y_2 - y_1 = m(x_2 - x_1) \rightarrow y - -1 = \frac{-3}{2}(x - -2)$ **[1]**

$3x + 2y + 8 = 0$ **[1]**

3. a) $\left(9, \frac{5}{4}\right)$ **[1]**

b) i) Know line passes through origin, since $y = kx + 0$, and obtain $k = \frac{5}{12}$ **[1]**

ii) Use substitution with the simultaneous equations to get a quadratic in x **[1]**

P has coordinates (12, 5) **[1]**

Alternate methods include using congruent triangles

iii) Gradient of tangent $= \frac{-12}{5}$ **[1]**

Substitute values for gradient and P into an equation and attempt to rearrange **[1]**

$12x + 5y - 169 = 0$ **[1]**

Binomial Expansion

QUICK TEST (page 31)

1. $27x^3 - 54x^2 + 36x - 8$
2. 4368
3. −2048
4. x^5 is 462 and x^6 is −462
5. $(2)^3 + 3\times(2)^2\times(-\sqrt{3}) + 3\times(2)(-\sqrt{3})^2 + (-\sqrt{3})^3 \rightarrow 26 - 15\sqrt{3}$
6. 160
7. $\frac{-99a^5}{16}$
8. $a = 2$

PRACTICE QUESTIONS (page 31)

1. Use of ${}^nC_r = \binom{n}{r} = \frac{n!}{r!(n-r)!}$ to find 243, 810, 1080 as the coefficients, getting at least two correct **[1]**

For an expression $ax^5 + bx^3 + cx + \cdots$ for getting at least two correct **[1]**

For the fully correct answer $243x^5 + 810x^3 + 1080x + \cdots$ **[1]**

The full working would look something like this:

$$\left(3x + \frac{2}{x}\right)^5 = 1\times(3x)^5 + 5(3x)^4\times\left(\frac{2}{x}\right)^1 + {}^5C_2(3x)^3\times\left(\frac{2}{x}\right)^2 + \cdots$$
$$= 3^5x^5 + 5\times3^4\times2\times x^3 + 10\times3^3\times2^2\times x + \ldots$$
$$= 243x^5 + 810x^3 + 1080x + \ldots$$

2. a) i) $a = \sqrt[4]{2401} = \pm7$

Since $a < 5$ the value of a is −7 **[1]**

ii) $f(x) = \frac{-2744x}{{}^4C_1(-7)^3} = \frac{2744x}{1372}$ **[1]**

$= 2x$ **[1]**

b) $q = (2)^4 = 16$ **[1]**

$p = {}^4C_3(-7)^1(2)^3 = -224$ **[1]**

c) $(2a + f(x))^4 = 2401\times2^4 - 2744\times2^3x + 1176\times2^2x^2 - 224\times2x^3 + 16x^4$ **[1]**

$= 38416 - 21952x + 4704x^2 - 448x^3 + 16x^4$ **[1]**

$\frac{d}{dx}(2a + f(x))^4 = -21952 + 9408x - 1344x^2 + 64x^3$ **[2]**

3. $10\times\left({}^4C_3\times(2)^3\times\left(-\frac{1}{2}\right)\right) = -160$ **[1]**

${}^5C_2\times a^2\times(-1)^3 = -160$ **[1]**

$10a^2 = 160$ **[1]**

$a^2 = 16$

$a = 4$ **[1]**

4. a) ${}^nC_2 = \frac{n\times(n-1)}{2}$ **[1]**

b) ${}^nC_2\times3^2\times1^{n-2} = 252$ **[1]**

$\frac{n^2-n}{2}\times9\times1 = 252$

$n^2 - n = 56$

$n^2 - n - 56 = 0$ **[1]**

$(n+7)(n-8) = 0$

$n = -7$, not relevant as n is positive **[1]**

$n = 8$ **[1]**

Trigonometry 1

QUICK TEST (page 35)

1. 8.14 (3 s.f.)
2. a) 4.23 cm (3 s.f.) b) 8.51 (3 s.f.) c) $9\sqrt{3} = 15.6$ (3 s.f.)
3. a) 38.7° (3 s.f.) b) 36.5° (3 s.f.) c) 90°
4. 8

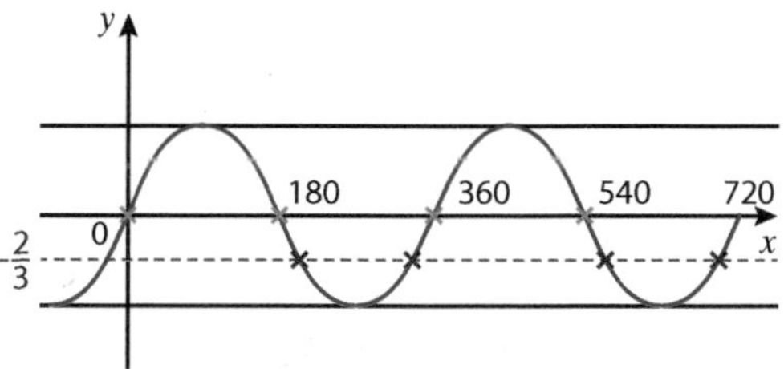

5. $7\sqrt{6} + 7\sqrt{2}$

PRACTICE QUESTIONS (page 35)

1. a) For factorising the quadratic (or using another method to solve the quadratic) $2(2c-1)(c+3)$ **[1]**

Both $\cos\alpha = \frac{1}{2}$ and $\cos\alpha = -3$ seen, dismissing −3 as impossible value, leading to $\cos\alpha = \frac{1}{2}$ **[1]**

b) For $\alpha = \cos^{-1}\left(\frac{1}{2}\right) = 60$ **[1]**

$\alpha = -60°, 60°, 300°$ **[1]**

c) For results from part **b)**, $\alpha = -60°, 60°, 300°$ **[1]**

For recognising $\cos\alpha = 0$ yields results **[1]**

For a full set of results: $\alpha = -90°, -60°, 60°, 90°, 270°, 300°$ **[1]**

2. $\tan x = 0$ or $3\cos x - 1 = 0$ or $5\sin x - 2 = 0$ **[1]**

$\tan x = 0 \Rightarrow x = -180°, 0°, 180°, 360°$ **[1]**

$3\cos x - 1 = 0 \Rightarrow \cos x = \frac{1}{3} \Rightarrow x = -70.53°, 70.53°, 289.47°$ **[1]**

$5\sin x - 2 = 0 \Rightarrow \sin x = \frac{2}{5} \Rightarrow x = 23.58, 156.42°$ **[1]**

−180°, −70.53°, 0°, 23.58°, 70.53°, 156.42°, 180°, 289.47°, 360°
(all answers to 2 d.p.)

3. a) Identify which angles and sides belong together; can be in the form of a diagram or implied through working **[1]**

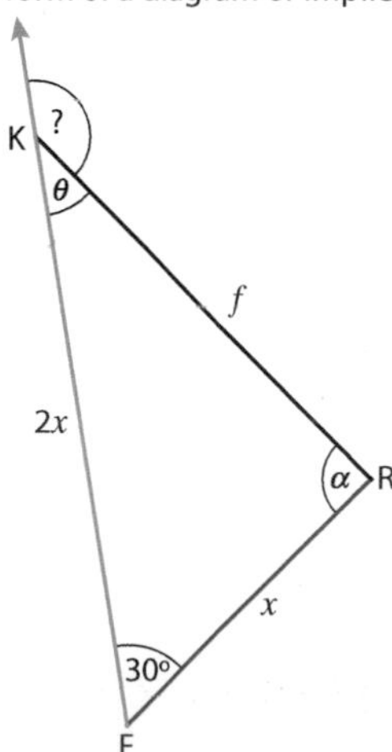

$f^2 = x^2 + (2x)^2 - 2 \times 2x \times x \times \cos 30$

For attempt to substitute into correct cosine rule **[1]**

$f^2 = 5x^2 - 2x^2 \times \sqrt{3} = (5 - 2\sqrt{3})x^2$

For finding the square of the distance between them **[1]**

$f = \sqrt{f^2}$

$f = \left[\sqrt{(5-2\sqrt{3})}\right] x = (5-2\sqrt{3})^{\frac{1}{2}} x$

For a correct answer in exact form **[1]**

b) Correctly recalling and selecting values to use in the sine rule (alternative method using cosine is possible) **[1]**

$\frac{\sin(30)}{(5-2\sqrt{3})^{\frac{1}{2}} x} = \frac{\sin(\theta)}{x} \rightarrow \sin(\theta) = \frac{\sin(30)}{(5-2\sqrt{3})^{\frac{1}{2}}} = 0.40344\ldots$ **[1]**

$\theta = 23.79397\ldots$ **[1]**

Angle which Kat has to look back through $= 180 - \theta$

$= 156.2060\ldots = 156°$ (3 s. f.) **[1]**

c) Gives any correct reason why the model wouldn't be realistic: the paths are unlikely to follow perfectly straight lines, it would be impossible for them to maintain the ratio of the speeds, etc. **[1]**

Trigonometry 2

QUICK TEST (page 39)

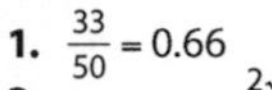

1. $\frac{33}{50} = 0.66$

2. 59.0°, 239.0°

3. $\cos\alpha = -\frac{2\sqrt{10}}{7}$

4. $\tan^{-1} 1 = 45°$

5. 8

6. 55.2°, 113.6°, 246.4°, 304.8°

7.

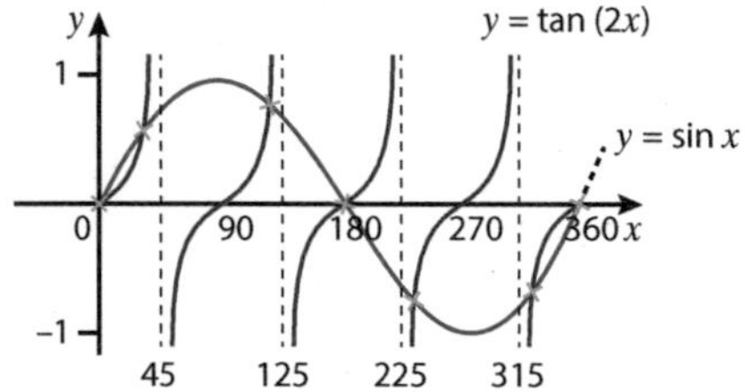

8. a) She hasn't noticed that the unknown angles are different, which means they can't be cancelled and simplified.

b) By cancelling by $\sin x$ she has lost a result where $\sin x = 0$

PRACTICE QUESTIONS (page 39)

1. a) $\sin x\,(12\sin x \tan x - 5) = 10\tan x$

$12\sin x \tan x - 5 = \frac{10}{\cos x}$

For cancelling through by common factor of $\sin x$ **[1]**

$12\sin^2 x - 5\cos x = 10$

For using $\tan x = \frac{\sin x}{\cos x}$ **[1]**

$12(1 - \cos^2 x) - 5\cos x = 10$

For using $\sin^2\theta + \cos^2\theta \equiv 1$ **[1]**

$12\cos^2 x + 5\cos x - 2 = 0$

$a = 12, b = 5, c = -2$

For correctly rearranging and finding values of a, b and c **[1]**

b) $\sin x = 0$ would also lead to some results to the original equation so Imran is incorrect **[1]**

c) $12\cos^2 x + 5\cos x - 2 = 0$

$(4\cos x - 1)(3\cos x + 2) = 0$ and $\sin x = 0$ **[1]**

$\cos x = \frac{1}{4}$, $\cos x = \frac{-2}{3}$, $\sin x = 0$ **[1]**

$\cos x = \frac{1}{4} \rightarrow x = 75.52248\ldots°, -75.52248\ldots°$ **[1]**

$\cos x = \frac{-2}{3} \rightarrow x = 131.8103\ldots°, -131.8103\ldots°$ **[1]**

$\sin x = 0 \rightarrow x = 0°, 180°$ **[1]**

2. a) $\sin x = \frac{5}{k}$

For there to be no solutions

$\frac{5}{k} < -1$ or $\frac{5}{k} > 1$ **[1]**

$-5 < k < 5$ **[1]**

b) $(2\sin x + 1)(k\sin x - 5) = 0$

So $a = 1$ and $b = -5$ **[2]**

c) $2\sin x + 1 = 0 \rightarrow \sin x = -\frac{1}{2} \rightarrow x = -30° \rightarrow x = 210°, 330°$ **[2]**

$k\sin x - 5 = 0 \rightarrow \sin x = \frac{5}{k}$ to have one result only $k = -5$ **[1]**

$\rightarrow x = 270°$ **[1]**

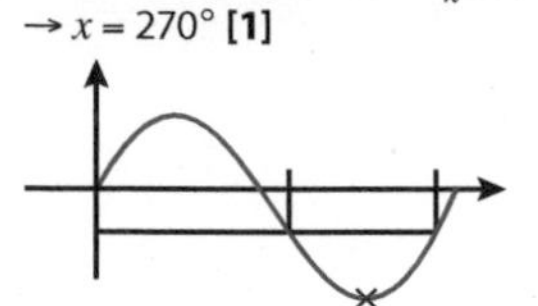

3. $24\left(1 - \sin^2\left(\frac{\theta}{2}\right)\right) = 2\sin\left(\frac{\theta}{2}\right) + 19$ **[1]**

$24 - 24\sin^2\left(\frac{\theta}{2}\right) = 2\sin\left(\frac{\theta}{2}\right) + 19 \rightarrow 24\sin^2\left(\frac{\theta}{2}\right) + 2\sin\left(\frac{\theta}{2}\right) - 5 = 0$ **[2]**

$\sin\left(\frac{\theta}{2}\right) = \frac{5}{12}$ **[1]**

$\sin\left(\frac{\theta}{2}\right) = -\frac{1}{2}$ **[1]**

$\theta = -670.8°, -409.2°, 49.2°, 310.8°$ **[1]**

$\theta = -300°, -60°, 420°, 660°$ **[1]**

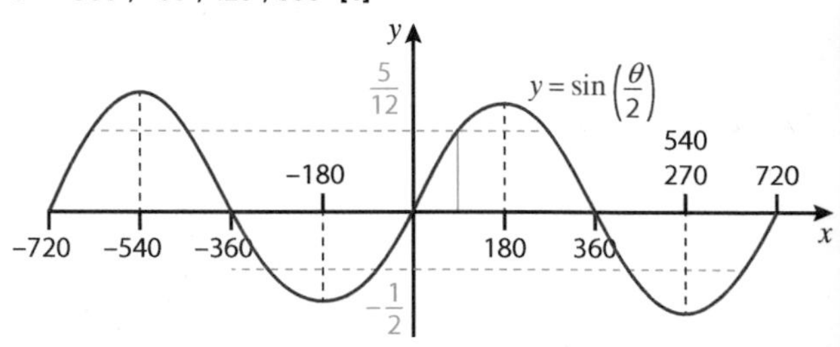

Day 3

Exponentials

QUICK TEST (page 43)

1. $4^{-2} = \frac{1}{16} = 0.0625$

2. and 3.

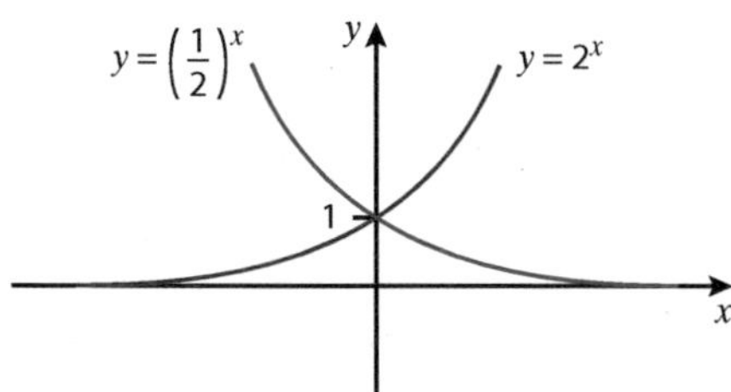

4. Reflection in the y-axis

5.

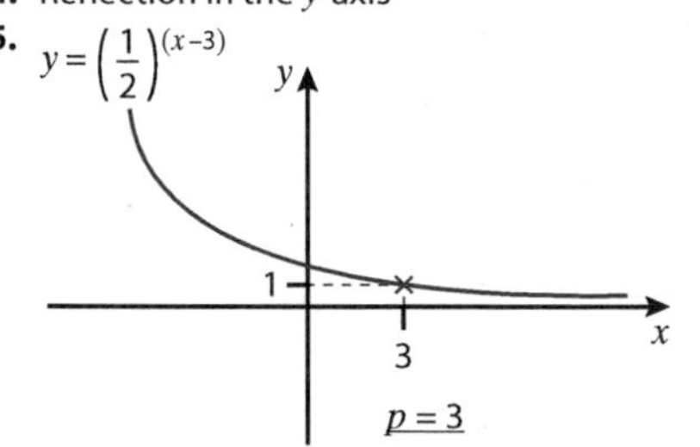

6. $\frac{1}{e} = 0.367\,87\ldots$

7. a) As the coefficient of t is positive, the population is increasing.

b) 34 000

c) 41 527 (rounded down to nearest person)

PRACTICE QUESTIONS (page 43)

1. a) $a = -1$ **[1]**

b) Stretch with factor 64 parallel to y-axis. **[1]**

$64 \times 4^x = 4^3 \times 4^x = 4^{x+3}$ **[1]**

Translation $\begin{pmatrix} -3 \\ 0 \end{pmatrix}$ **[1]**

c) 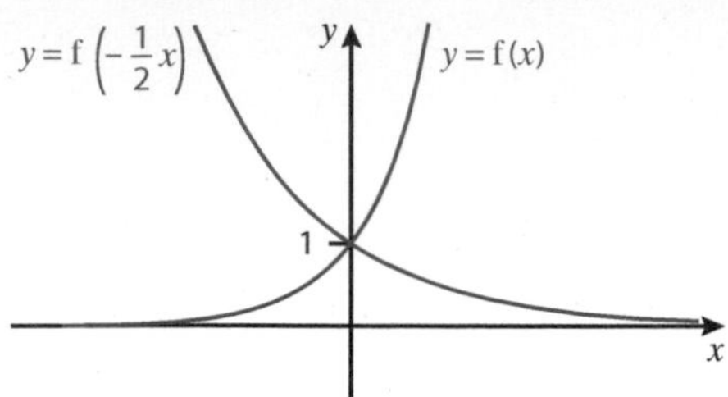

For graph of $y = f(x)$ with y-intercept labelled **[1]**
For reflection in y-axis and y-intercept at 1 **[1]**
For a reasonable attempt at showing a stretch of scale factor 2 in the x-direction **[1]**

2. a) 3200 **[1]**
b) Increase is $0.0327 \times 3200e^{0.0327 \times 10}$ **[1]**
= 145 tigers per year (to the nearest whole tiger) **[1]**
c) $3200e^{0.0327 \times 44}$ **[1]**
= 13 490 (to the nearest whole tiger) **[1]**

Logarithms

QUICK TEST (page 47)

1. $x = 4$
2. 7
3. 4
4. $q = \log_a (b - 5) + 2$
5. 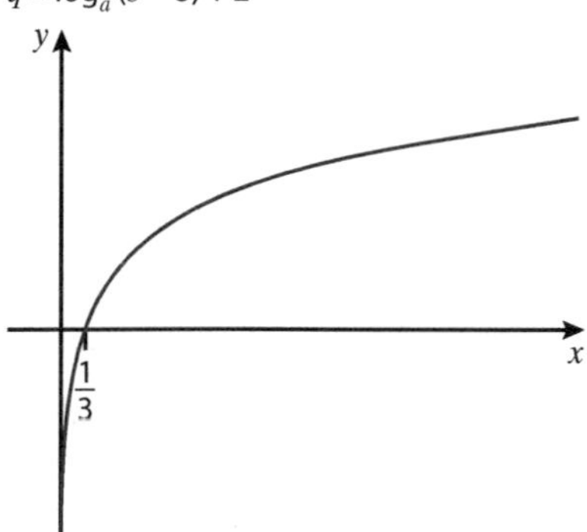

6. –3
7. 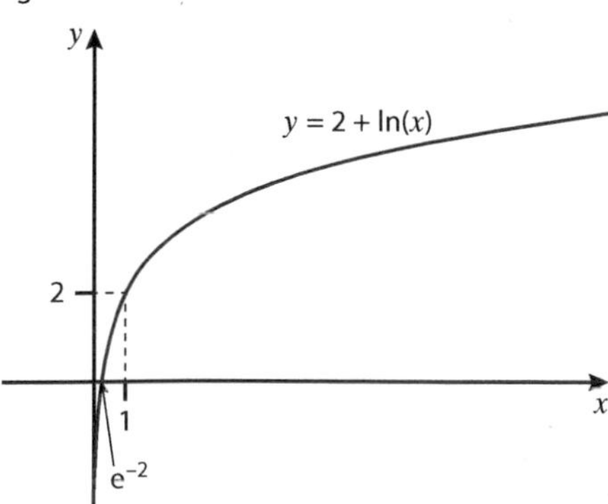

8. 2
9. ln 7 + 1
10. 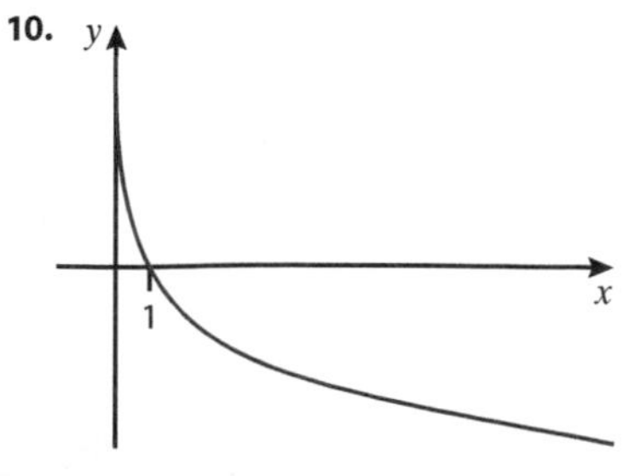

PRACTICE QUESTIONS (page 47)

1. a) $b = 4^a$ **[1]**
b) $b = (2^2)^a = 2^{2a}$ **[2]**
$\log_2 b = 2a$ **[1]**
2. a) $p = 2$ **[1]**
b) 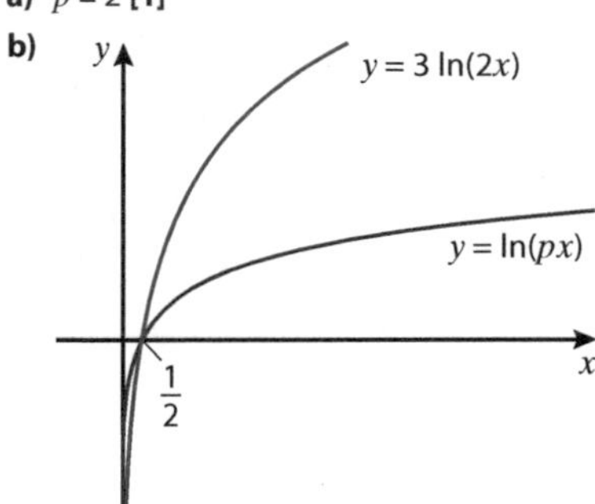

[2]
c) $-15 = 3\ln(2x)$ Substituting in x-value **[1]**
$\ln(2x) = -5$ **[1]**
$e^{\ln(2x)} = e^{-5}$ **[1]**
$2x = e^{-5} \rightarrow x = \frac{e^{-5}}{2} = \frac{1}{2e^5}$ **[1]**
3. $c = 25$ **[1]**
$25 = 3e^{2x+1} - 5$ **[1]**
$30 = 3e^{2x+1} \rightarrow e^{2x+1} = 10$ For rearrangement **[1]**
$2x + 1 = \ln 10$ For use of natural logarithm **[1]**
$2x = \ln(10) - 1$
$x = \frac{1}{2}(\ln(10) - 1)$ For further rearrangement **[1]**
$a = \frac{1}{2}, b = -\frac{1}{2}$ and $c = 25$ **[1]**

Log Rules

QUICK TEST (page 51)

1. $2^7 = 128$
2. $x = \log_4 63$
3. a) 12 b) 2 c) 5
d) 3 e) $4^5 = 1024$ f) 6
4. a) $\frac{\log_3 12}{\log_3 9} = \frac{\log_3 12}{2} = \frac{1}{2}\log_3 12$
b) $\frac{\log_4 15}{\log_4 2} = \frac{\log_4 15}{\frac{1}{2}} = 2\log_4 15$
c) $\frac{\log_2 4}{\log_2 e} = \frac{2}{\log_2 e}$
5. $3 + \log_2 3 \rightarrow a = 3$
6. $\log_a b + \frac{1}{2}\log_a c$

PRACTICE QUESTIONS (page 51)

1. $\frac{\log_3 12}{\log_3 9} = \frac{\log_3 12}{2} = \frac{1}{2}\log_3 12$ **[1]**
$= \frac{1}{2}(\log_3 (3 \times 4)) = \frac{1}{2}(\log_3 3 + \log_3 4)$ **[1]**
$= \frac{1}{2}(1 + \log_3 4)$ **[1]**
2. a) $(3^x - 12)(3^x - 3) = 3^{2x} - 12 \times 3^x - 3 \times 3^x + 36$
For expanding the brackets **[1]**
$= 3^{2x} - 4 \times 3 \times 3^x - 3 \times 3^x + 36 = 3^{2x} - 4 \times 3^{x+1} - 3^{x+1} + 36$
For using coefficients to create 3^{x+1} term **[1]**
For simplification and final answer with fully convincing, supporting steps **[1]**
b) $3^{4x} - 5 \times 3^{2x+1} + 36 = (3^{2x} - 12)(3^{2x} - 3)$
For using factorisation from part **a)** but with 3^{2x} **[1]**
$(3^{2x} - 12)(3^{2x} - 3) = 0 \rightarrow 3^{2x} = 3$ and $3^{2x} = 12$ **[1]**
$3^{2x} = 3 \rightarrow 2x = \log_3 3 = 1 \rightarrow x = \frac{1}{2}$ **[1]**
$3^{2x} = 12 \rightarrow 2x = \log_3 12 = 1 + \log_3 4 = 1 + \log_3 (2^2) \rightarrow x = \frac{1}{2} + \log_3 2$ **[1]**

3. $4 = \log_2(x + 13) + \log_2(x - 2)$ or $\log_2(x + 13) + \log_2(x - 2) - \log_2 16 = 0$ **[1]**
For correctly combining the logs
$4 = \log_2(x + 13)(x - 2)$ or $\log_2 \frac{(x+13)(x-2)}{16} = 0$ **[1]**
For rewriting in index form (and expanding brackets)
$2^4 = x^2 + 11x - 26$ or $\frac{x^2 + 11x - 26}{16} = 2^0$ **[1]**
For evaluating and rearranging into quadratic to solve
$16 = x^2 + 11x - 26 \quad x^2 + 11x - 26 = 1 \times 16$
$x^2 + 11x - 42 = 0$ **[1]**
For solving the quadratic using any valid method to give $x = -14, x = 3$ **[1]**
For discounting results that would give a negative within the log in the original question leaving $x = 3$ **[1]**

Using Logarithms

QUICK TEST (page 55)

1. 3.47 (3 s.f.)
2. $Y = \log y$, $X = \log x$ and $c = \log a$
3. log 6
4. a) $k = 1000$ b) $b = \sqrt{10}$ c) $y = 100000$
5. a) 200 b) 157.0 g (1 d.p.)

PRACTICE QUESTIONS (page 55)

1. a) For setting up equation with initial conditions:
$100 = Ae^{-k \times 0} + 12$ **[1]**
For rearranging to find A: $A = 100 - 12 = 88$ **[1]**
b) Using their A, substitute values in for 4 minutes:
$T = Ae^{-kt} + C \rightarrow 89 = 88e^{-k\times4} + 12$ **[1]**
For rearranging to solve: $89 - 12 = 88e^{-k\times4}$
$e^{-k\times4} = \frac{77}{88} = \frac{7}{8}$ **[1]**
$-k \times 4 = \ln\frac{7}{8}$
$k = -\frac{1}{4}\ln\frac{7}{8} = \ln\left(\frac{8}{7}\right)^{\frac{1}{4}}$
For getting an answer in log form **[1]**
c) For setting up the inequality, or equation, to find key point: $t < 25$
$88e^{-t\ln\left(\frac{8}{7}\right)^{\frac{1}{4}}} + 12 = 25$ **[1]**
$88e^{\ln\left(\frac{8}{7}\right)^{\frac{-t}{4}}} = 13$
$e^{\ln\left(\frac{8}{7}\right)^{\frac{-t}{4}}} = \frac{13}{88}$ **[1]**
$\left(\frac{8}{7}\right)^{\frac{-t}{4}} = \frac{13}{88}$ **[1]**
$-\frac{t}{4} = \log_{\frac{8}{7}}\frac{13}{88}$ **[1]**
$t = -4\log_{\frac{8}{7}}\frac{13}{88} = 57.2865\ldots$ **[1]**
After 57.3 minutes (3 s.f.) the water will have cooled to 25 °C, $t > 57.3$ **[1]**
2. a) $\log_2 y = \log_2 ax^n$ **[1]**
$\log_2 y = \log_2 x^n + \log_2 a$
$\log_2 y = n\log_2 x + \log_2 a$ **[1]**
$Y = nX + c$, where $Y = \log_2 y$, $X = \log_2 x$ and $c = \log_2 a$ **[1]**
b) n = gradient = 2.5 **[1]**
c = y-intercept = 3 **[1]**
$c = \log_2 a$ **[1]**
$a = 2^3 = 8$ **[1]**
c) $y = 8 \times x^{2.5} = 8 \times 100^{2.5}$ **[1]**
$y = 800000$ **[1]**

Day 4

Differentiation 1

QUICK TEST (page 59)

1. a) $6x$ b) 5 c) $\frac{4}{5}x^{-\frac{1}{5}}$
d) $-\frac{3}{2}x^{-\frac{3}{2}}$ e) $-12x^{-4}$ f) 3
g) $20x^3 + 9x^2$
2. a) $2^3 + 3\times2^2\times x^{\frac{1}{2}} + 3\times2\times x + x^{\frac{3}{2}} = 8 + 12x^{\frac{1}{2}} + 6x + x^{\frac{3}{2}}$
b) Hence $\frac{dy}{dx} = 6x^{-\frac{1}{2}} + 6 + \frac{3}{2}x^{\frac{1}{2}}$
c) Gradient $= \frac{dy}{dx} = \frac{6}{\pm2} + 6 + \frac{3}{2}\times\pm2 \rightarrow 12$ and 0
3. a) $\frac{dy}{dx} = -8x^{-3} + 1$ b) $\frac{dy}{dx} = 0$

PRACTICE QUESTIONS (page 59)

1. a) For rewriting in index form $\left(3x^{\frac{1}{2}} + 2x^2\right)^3$ **[1]**
For first step in expanding brackets using any correct method i.e. binomial expansion
$y = \left(3x^{\frac{1}{2}}\right)^3 + 3\left(3x^{\frac{1}{2}}\right)^2(2x^2) + 3\left(3x^{\frac{1}{2}}\right)(2x^2)^2 + (2x^2)^3$ **[1]**
For simplifying each term $y = 27x^{\frac{3}{2}} + 54x^3 + 36x^{\frac{9}{2}} + 8x^6$ **[2]**
b) $\frac{dy}{dx} = \left(27\times\frac{3}{2}\right)x^{\frac{1}{2}} + (54\times3)x^2 + \left(36\times\frac{9}{2}\right)x^{\frac{7}{2}} + (8\times6)x^5$
For correct indices **[1]**
$\frac{dy}{dx} = \frac{81}{2}x^{\frac{1}{2}} + 162x^2 + 162x^{\frac{7}{2}} + 48x^5$
For correct coefficients **[1]**
c) For clear attempt to substitute in x-value to $\frac{dy}{dx}$
$\frac{81}{2}\times(4)^{\frac{1}{2}} + 162\times(4)^2 + \cdots$ **[1]**
72561 **[1]**
If using a calculator, both marks are awarded for the correct final answer. However, without showing the substitution, if a small error is made on the calculator it will lose both marks.
2. a)

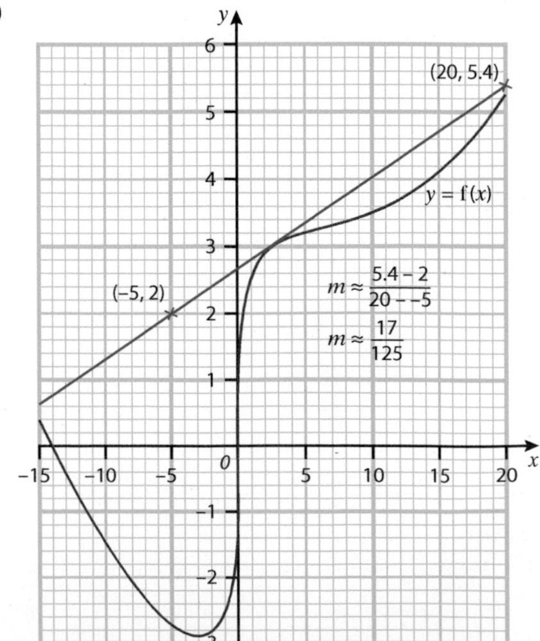

For a reasonable line of best fit drawn touching the curve at $x = 3$ **[1]**

For a suitable pair of coordinates identified and read correctly with a small margin of error **[1]**

For the gradient being found correctly from the pair of coordinates **[1]**

Note: The answer shown is an approximation. The exact same answer is highly unlikely, but your answer should be similar.

b)

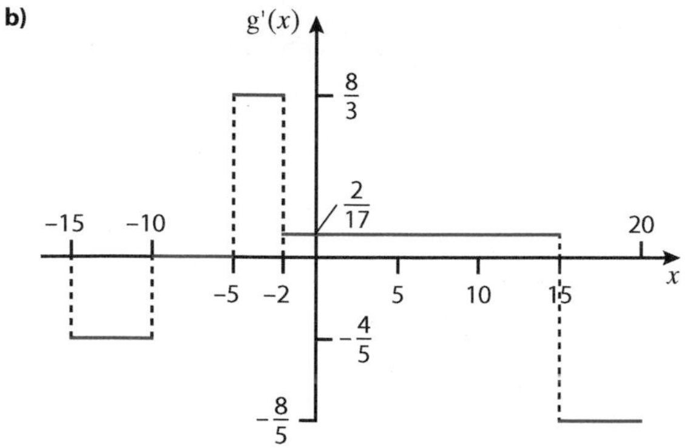

For a set of horizontal lines, correctly placed above, below or on the axis **[1]**

For the correct values labelled clearly on the y-axis (lose 1 mark for each mistake) **[2]**

Lose 1 mark for solid vertical lines (no vertical lines shown is okay as long as start and end points are clearly not overlapping)

Differentiation 2

QUICK TEST (page 63)

1. **a)** Increasing **b)** Stationary **c)** Decreasing **d)** Stationary **e)** Decreasing
2. **a)** $\frac{d^2y}{dx^2} = 2$ **b)** $\frac{d^2y}{dx^2} = 12x - 4$
3. The stationary points are: A minimum at (0, –4) and a maximum at (2, 0)
4. **a)** Gradient = –3 **b)** $a = -3$
5. Gradient $= -\frac{8}{7}$

PRACTICE QUESTIONS (page 63)

1. **a)** $f'(x) = 2x^2 + 8x + p$ For correct differentiation **[1]**

$2x^2 + 8x + p = 0$ For setting gradient function equal to 0 **[1]**

$b^2 - 4ac = 0$ For use of discriminant **[1]**

$8^2 - 4 \times 2 \times p = 0$

$64 - 8p = 0$ For correct substitution and rearrangement **[1]**

b) $f'(x) = 2x^2 + 8x + 8$ For finding p and using in gradient function **[1]**

$x^2 + 4x + 4 = 0$

$(x + 2)^2 = 0$ For valid method used for solving the quadratic **[1]**

$x = -2$ For correct x-value **[1]**

$y = f(-2) = \frac{2}{3}(-2)^3 + 4(-2)^2 + 8(-2) - 2$

$y = -\frac{22}{3} = -7\frac{1}{3}$

$\left(-2, -\frac{22}{3}\right)$ For correct y-coordinate for stationary point **[1]**

c) $f'(x) = 2x^2 + 8x + 6$

$2x^2 + 8x + 6 = 0$ For equating the differentiated equation to 0 **[1]**

$2(x^2 + 4x + 3) = 2(x + 1)(x + 3) = 0$

$x = -1$ and $x = -3$ For solutions for x **[1]**

$f''(x) = 4x + 8$ For finding the second differential **[1]**

$f''(-1) = -4 + 8 = 4$ This is positive so a minimum at $x = -1$ **[1]**

$f''(-3) = -12 + 8 = -4$ This is negative so a maximum at $x = -3$ **[1]**

$f(-1) = \frac{2}{3}(-1)^3 + 4(-1)^2 + 6(-1) - 2 = -4\frac{2}{3}$

$f(-3) = \frac{2}{3}(-3)^3 + 4(-3)^2 + 6(-3) - 2 = -2$

For getting: Minimum at $\left(-1, -4\frac{2}{3}\right)$ and maximum at (–3, –2) **[1]**

2. **a)** $y = 4(-1)^3 + 6(-1)^2 + (-1)$

P is at (–1, 1) [1]

$\frac{dy}{dx} = 12x^2 + 12x + 1$ For correct differentiation **[1]**

Gradient of curve at $x = -1$ is

$m_t = 12(-1)^2 + 12(-1) + 1 = 1$ **[1]**

$m_n = \frac{-1}{1} = -1$ For finding the gradient of the normal **[1]**

Normal equation

$y = -x$ For equation of the normal line **[1]**

$4x^3 + 6x^2 + x = -x$ For equating curve and normal **[1]**

$4x^3 + 6x^2 + 2x = 0$

$2x(x + 1)(2x + 1) = 0$ For method of solving cubic **[1]**

$2x = 0 \rightarrow x = 0 \rightarrow y = 0$ **R is at (0, 0) [1]**

$2x + 1 = 0 \rightarrow x = -0.5 \rightarrow y = 0.5$ **Q is at** $\left(-\frac{1}{2}, \frac{1}{2}\right)$ **[1]**

b) Gradient at $x = -0.5$

$12(-0.5)^2 + 12(-0.5) + 1$

Gradient of curve at Q = –2 ∴ function is decreasing **[1]**

Integration 1

QUICK TEST (page 67)

1. **a)** $x^3 + C$ **b)** $\frac{5}{2}x^2 + C$ **c)** $\frac{5}{9}x^{\frac{9}{5}} + C$ **d)** $6x^{\frac{1}{2}} + C$ **e)** $-2x^{-2} + C$ **f)** $\frac{3}{2}x^2 + 2x + C$ **g)** $x^5 + \frac{3}{4}x^4 - 3x + C$
2. **a)** $f(x) = 3x^{\frac{7}{3}} + 2x^{-1} + C$ **b)** $C = 15.75$
3. **a)** $v = \int 3t + 2\,dt = \frac{3}{2}t^2 + 2t + C$

b) $10 = \frac{3}{2}(0)^2 + 2 \times 0 + C$

$C = 10$

PRACTICE QUESTIONS (page 67)

1. **a)** When $x = 0.8$ $\frac{dy}{dx} = 10 \times 0.8 + 2 = 10$ **[1]**

Since the gradient function forms a tangent to the curve at this point, the quadratic also passes through (0.8, 10) **[1]**

b) $y = \int \frac{dy}{dx}\,dx = \int 10x + 2dx = 5x^2 + 2x + C$ **[1]**

$10 = 5 \times (0.8)^2 + 2 \times 0.8 + C$ **[1]**

$c = 10 - 3.2 - 1.6 = 5.2$ **[1]**

$y = 5x^2 + 2x + 5.2$ **[1]**

2. **a)** $f(x) = \int ax^2 + 2x + a^2\,dx = \frac{a}{3}x^3 + x^2 + a^2x + C$ **[2]**

b) $f(0) = -7$ $-7 = \frac{a}{3} \times 0^3 + 0^2 + a^2 \times 0 + C$ **[1]**

$C = -7$ **[1]**

$f(1) = -4.25$ $-4.25 = \frac{a}{3} \times 1^3 + 1^2 + a^2 \times 1 - 7$ **[1]**

$-51 = 4a + 12 + 12a^2 - 84 \rightarrow 12a^2 + 4a - 21 = 0$ **[1]**

$a = \frac{-4 \pm \sqrt{4^2 - 4 \times 12 \times -21}}{24}$ **[1]**

$a = \frac{7}{6}$ or $a = -\frac{3}{2}$

As $a > 0$, $a = \frac{7}{6}$ **[1]**

3. a) $y=\left(px^{\frac{1}{2}}+2x\right)(x^q-1)$

For conversion to index notation **[1]**

$y=px^{q+\frac{1}{2}}+2x^{q+1}-px^{\frac{1}{2}}-2x$

For correct expansion of brackets **[1]**

b) $\int y\,dx=\frac{p}{q+\frac{3}{2}}x^{q+\frac{3}{2}}+\frac{2}{q+2}x^{q+2}-\frac{2p}{3}x^{\frac{3}{2}}-x^2+C$ **[4]**

[Lose 1 mark for each error]

c) Equate coefficients of $x^{\frac{3}{2}}$ term

$-\frac{2p}{3}=-8 \quad 2p=24 \quad p=12$ **[1]**

As p and q are positive $x^3=x^{q+\frac{3}{2}}$

$3=q+\frac{3}{2} \quad q=\frac{3}{2}$ **[1]**

Check all values work with these values

$\int y\,dx=\frac{12}{\frac{3}{2}+\frac{3}{2}}x^{\frac{3}{2}+\frac{3}{2}}+\frac{2}{\frac{3}{2}+2}x^{\frac{3}{2}+2}-\frac{2\times12}{3}x^{\frac{3}{2}}-x^2+C$

$=4x^3+\frac{4}{7}x^{\frac{7}{2}}-8x^{\frac{3}{2}}-x^2+C$ **[1]**

Integration 2

QUICK TEST (page 72)

1. **a)** 12 units^2 **b)** $5\sqrt{2}$ units^2 **c)** 15 units^2
2. **a)** Positive **b)** Negative
3. 24
4. 144
5. 45 units^2
6. $\int_{-2}^{0}2x^3-4x+7\,dx=\left[\frac{1}{2}x^4-2x^2+7x\right]_{-2}^{0}$

$=\left(\frac{1}{2}\times0^4-2\times0^2+7\times0\right)$

$-\left(\frac{1}{2}\times(-2)^4-2\times(-2)^2+7\times(-2)\right)$

$=(0)-(-14)=14$

PRACTICE QUESTIONS (page 73)

1. For factorisation or other correctly applied method to find roots:

$y=-(x-2)(2x+3)=(2-x)(2x+3)$ **[1]**

For establishing bounds as between roots found $\int_{-\frac{3}{2}}^{2}y\,dx$ **[1]**

$\int_{-\frac{3}{2}}^{2}-2x^2+x+6\,dx=\left[-\frac{2}{3}x^3+\frac{1}{2}x^2+6x\right]_{-\frac{3}{2}}^{2}$ **[1]**

$=\left(-\frac{2}{3}(2)^3+\frac{1}{2}(2)^2+6(2)\right)-\left(-\frac{2}{3}\left(-\frac{3}{2}\right)^3+\frac{1}{2}\left(-\frac{3}{2}\right)^2+6\left(-\frac{3}{2}\right)\right)$

For correct substitution of bounds **[1]**

Area $=\frac{343}{24}=14\frac{7}{24}=14.29$ (2 d.p.) **[1]**

2. a) $\int_{-3}^{1}x^3+12x^{\frac{4}{3}}+6x+6\,dx=\left[\frac{1}{4}x^4+\frac{36}{7}x^{\frac{7}{3}}+3x^2+6x\right]_{-3}^{1}$ **[2]**

$=\left(\frac{1}{4}\times1^4+\frac{36}{7}\times1^{\frac{7}{3}}+3\times1^2+6\times1\right)$

$-\left(\frac{1}{4}\times(-3)^4+\frac{36}{7}\times(-3)^{\frac{7}{3}}+3\times(-3)^2+6\times(-3)\right)$ **[1]**

$=\frac{403}{28}--37.505555...$

$=51.89840868...$ **[1]**

$=51.9$ (3 s.f.)

b)

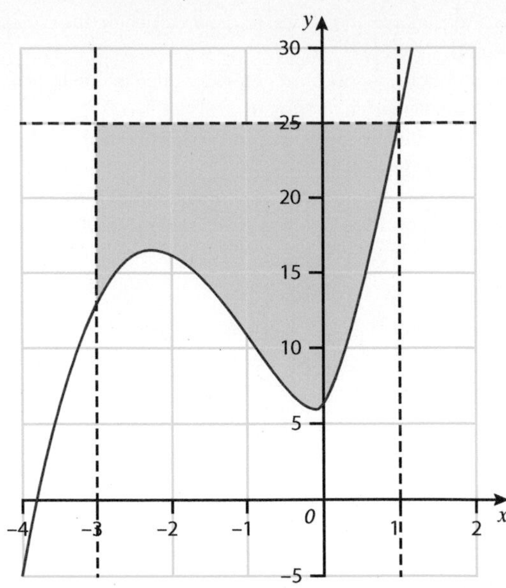

Area of rectangle = 25 × 4 = 100 **[1]**

100 – answer from part **a)** **[1]**

48.1 **[1]**

c)

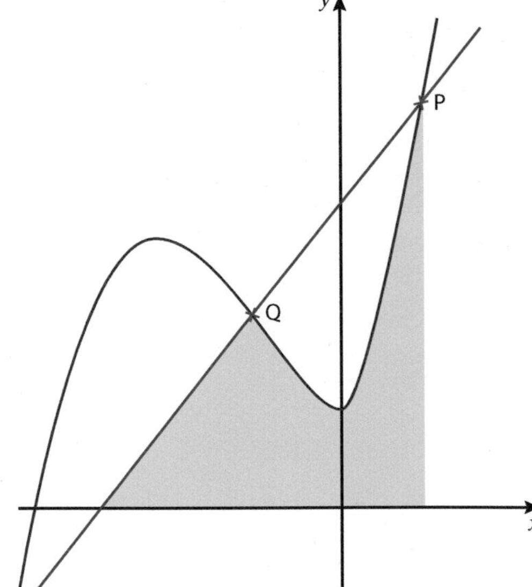

$x^3+12x^{\frac{4}{3}}+6x+6=7x+18$

For equating to find intersections **[1]**

$x^3+12x^{\frac{4}{3}}-x-12=0$

For rearranging **[1]**

d) Let $f(x)=x^3+12x^{\frac{4}{3}}-x-12$

$f(1)=1+12-1-12=0$

$\therefore x=1$ is a solution to the equation

$f(-1)=-1+12+1-12=0$

$\therefore x=-1$ is a solution to the equation **[1]**

At $x=1$, $y=7+18=25$

P is at (1, 25)

At $x=-1$, $y=-7+18=11$

Q is at (−1, 11) **[1]**

e) Area $=\int_{-1}^{1}y\,dx+$ area of triangle $\left(\frac{1}{2}bh\right)$ **[1]**

From part **d)** $h=11$

Line intersects x-axis at $7x = -18$ $\quad x = -\frac{18}{7}$ **[1]**

$b = -1 - \frac{-18}{7} = \frac{11}{7}$

Area of triangle $= \frac{1}{2} \times b \times h = \frac{1}{2} \times 11 \times \frac{11}{7} = \frac{121}{14}$ **[1]**

$$\int_{-1}^{1} y\,dx = \left[\frac{1}{4}x^4 + \frac{36}{7}x^{\frac{7}{3}} + 3x^2 + 6x\right]_{-1}^{1}$$

$$= \left(\frac{1}{4} \times 1^4 + \frac{36}{7} \times 1^{\frac{7}{3}} + 3 \times 1^2 + 6 \times 1\right)$$

$$- \left(\frac{1}{4} \times (-1)^4 + \frac{36}{7} \times (-1)^{\frac{7}{3}} + 3 \times (-1)^2 + 6 \times (-1)\right) \textbf{ [1]}$$

$$= \frac{403}{28} - \left(-\frac{221}{28}\right) = \frac{156}{7} \textbf{ [1]}$$

Shaded area $= \frac{156}{7} + \frac{121}{14} = \frac{433}{14} = 30.9$ (3 s.f.) **[1]**

Day 5

Modelling, Quantities and Units

QUICK TEST (page 77)

1. a) 2500 m b) 10 800 s c) 4200 kg
2. a) 340 N b) $\frac{1}{500} = 0.002\ \text{ms}^{-1}$ c) $\frac{1}{2} = 0.5\ \text{ms}^{-2}$
3. A–S, B–P, C–T, D–Q, E–U, F–R

Kinematics and Graphs

QUICK TEST (page 83)

1. Velocity
2. a) $\frac{5}{3} = 1.67\ \text{ms}^{-1}$ (3 s.f.)
   b) Highest speed section C = $2.5\ \text{ms}^{-1}$
   c) 12 s and at 22 s
   d) The object is stationary for 4 seconds
3. a) $0\ \text{ms}^{-2}$ b) $-\frac{5}{2} = -2.5\ \text{ms}^{-2}$
   c) $\frac{10}{3} = 3.33\ \text{ms}^{-2}$ (3 s.f.) d) 40 m
   e) $p = 15$
   f) $40 + 20 + \frac{1}{2}(10)(7) + \frac{1}{2}(9)(30) = 230$ m
   g) 40 + 20 – 35 + 135 = 160 m

PRACTICE QUESTIONS (page 83)

1. a) Metres per second to metres per hour × 3600
   18 × 3600 = 64 800 mh^{-1} **[1]**
   Metres per hour to miles per hour ÷ 1610
   64 800 ÷ 1610 = 40.2484 …
   The new speed limit is 40 mph. **[1]**
   b) Area under the graph = 80 × 13 = 1040 m **[1]**

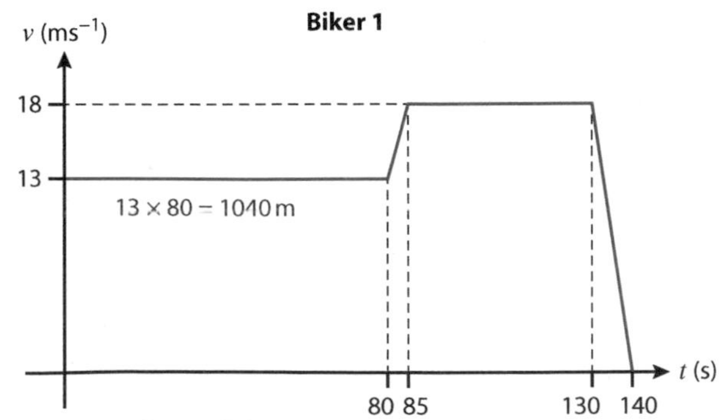

   c) B is 1040 m from A. **[1]**
   Area from $t = 0$ to $t = 10$:
   $\frac{1}{2} \times 13 \times 10 = 65$ m **[1]**
   1040 – 65 = 975 m

$(p - 10) \times 13 = 975$ **[1]**

$p - 10 = 75$

$p = 85$ s **[1]**

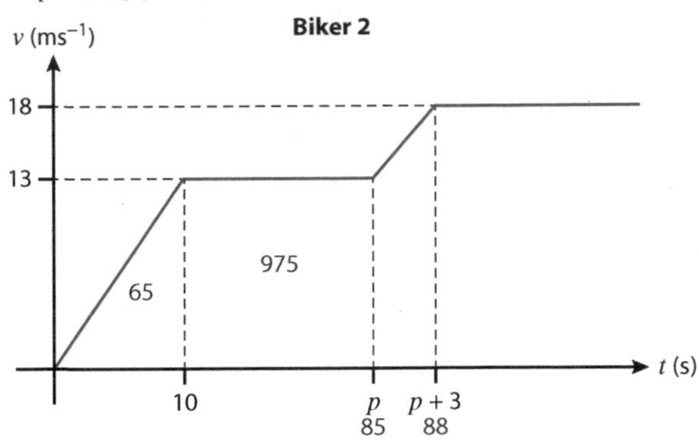

2. The question doesn't specify use of a velocity–time graph but it is generally a good idea to use one when faced with a multi-part journey.

   a)

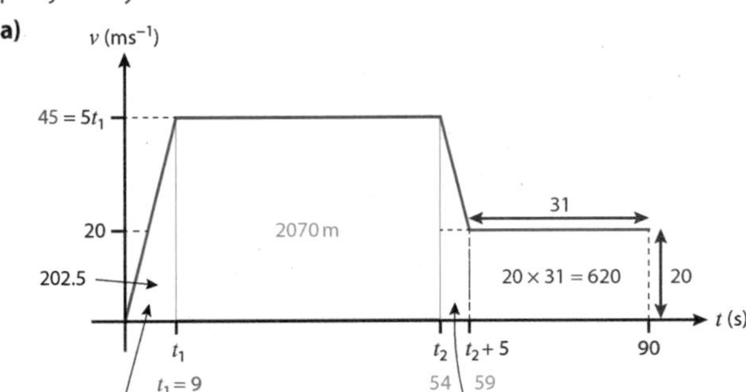

Area = displacement

$202.5 = \frac{1}{2} \times t_1 \times 5t_1$

$2.5t_1^2 = 202.5$

$t_1^2 = 81$

$t_1 = 9$ s (ignore negative)

$2070 = 45(t_2 - 9)$

$t_2 - 9 = 46$

$t_2 = 54$ s

Area $= \frac{1}{2}(45 + 20) \times 5 = 162.5$ m

Total distance PQ =
202.5 + 2070 + 162.5 + 620
= 3055 m

For using area of triangle equal to 202.5 **[1]**
For finding $t_1 = 9$ s **[1]**
For finding velocity 45 ms^{-1} **[1]**
For calculation 2070 ÷ '45' **[1]**
For finding $t_2 = 54$ s **[1]**
For finding area between 54 and 59 seconds (162.5 m) **[1]**
For finding total distance from P to Q (3055 m) **[1]**

   b)

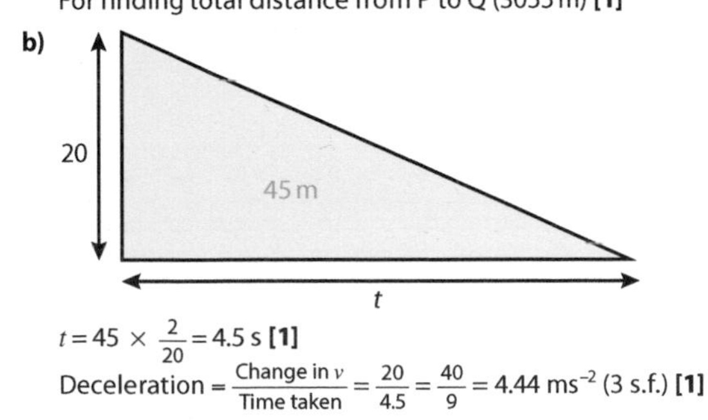

$t = 45 \times \frac{2}{20} = 4.5$ s **[1]**

Deceleration $= \frac{\text{Change in } v}{\text{Time taken}} = \frac{20}{4.5} = \frac{40}{9} = 4.44\ \text{ms}^{-2}$ (3 s.f.) **[1]**

Vectors 1

QUICK TEST (page 87)

1. tan 45 = 1
2. $\theta = 53.1°$

3. a) $\mathbf{p} = -2\mathbf{i} + \frac{3}{4}\mathbf{j}$

b)

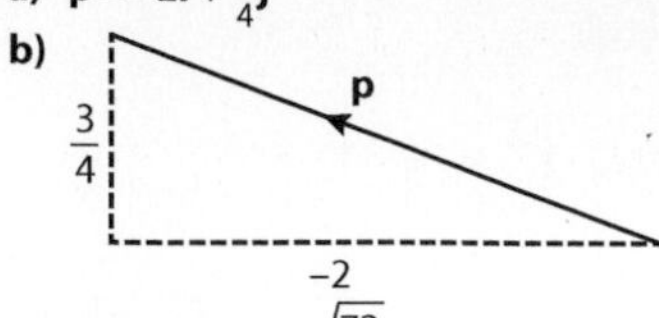

c) Magnitude $= \frac{\sqrt{73}}{4}$

d)

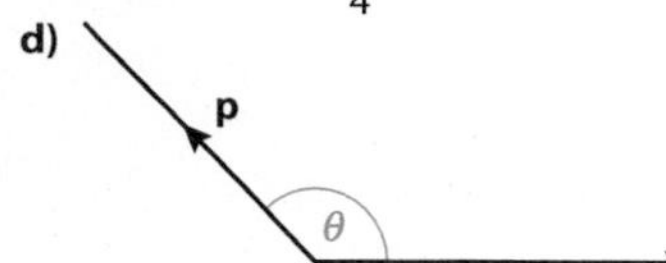

$\theta = 159.4°$ (1 d.p.)

4. $\mathbf{a} = -\mathbf{i} + \mathbf{j} = \begin{pmatrix} -1 \\ 1 \end{pmatrix}$

$\mathbf{b} = -\mathbf{i} - 2\mathbf{j} = \begin{pmatrix} -1 \\ -2 \end{pmatrix}$

$\mathbf{c} = \mathbf{i} + 2\mathbf{j} = \begin{pmatrix} 1 \\ 2 \end{pmatrix}$

5. $\begin{pmatrix} 8 \\ -9 \end{pmatrix}$

6. $3\mathbf{i} + 3\mathbf{j}$

7. $\begin{pmatrix} -425\sqrt{3} \\ -425 \end{pmatrix}$ kmh^{-1}

PRACTICE QUESTIONS (page 87)

1. a) Horizontally $2 - 7 = -5$

Vertically $10 - -2 = 12$ **[1]**

$\underset{AB}{\rightarrow} = \begin{pmatrix} -5 \\ 12 \end{pmatrix}$ **[1]**

b) $|\overrightarrow{AB}| = \sqrt{(-5)^2 + (12)^2}$ **[1]**

$= \sqrt{169} = 13$ **[1]**

c)

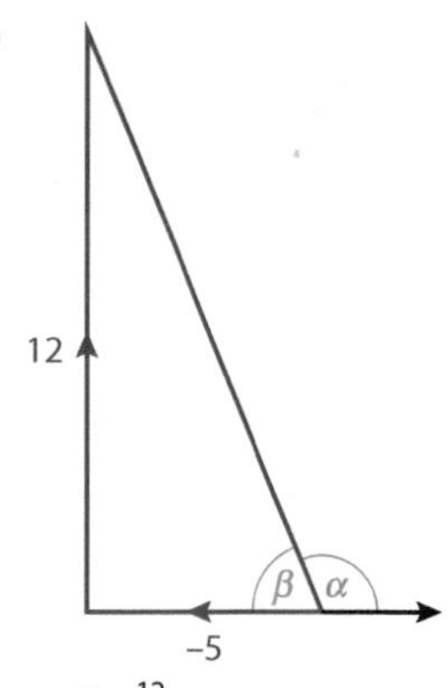

$\tan\beta = \frac{12}{5}$

$\beta = \tan^{-1}\frac{12}{5} = 67.3801\ldots$ **[1]**

$\alpha = 180 - 67.380\ldots = 112.6°$ (1 d. p.) **[1]**

2.

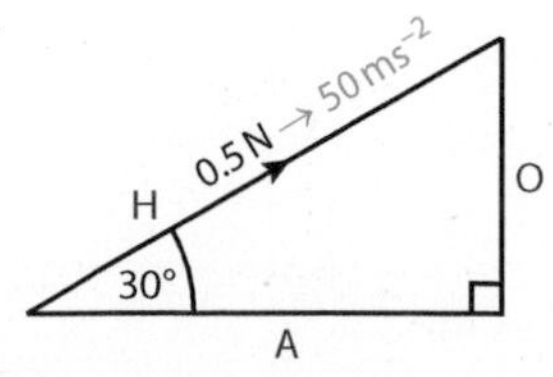

$a = \frac{0.5}{0.01} = 50$ ms^{-2} in the direction of the force **[1]**

$\text{O} = \text{H} \sin\theta = 50\sin30$ **[1]**

$q = 25$

$A = \text{H}\cos\theta = 50\cos30$ **[1]**

$p = 25\sqrt{3}$

For both p and q **[1]**

3. a) Distance travelled BC is $\sqrt{6^2 + 8^2} = \sqrt{100} = 10$

So distance AB is $23 - 10 = 13$ **[1]**

$(2a)^2 + (11 - a)^2 = 13^2$ **[1]**

$4a^2 + 121 - 22a + a^2 = 169$

$5a^2 - 22a - 48 = 0$ **[1]**

$a = 6$ (or $-\frac{8}{5}$ but ignore as question states that a is positive) **[1]**

$\overrightarrow{AC} = (2a + 6)\mathbf{i} + (11 - a - 8)\mathbf{j}$ km **[1]**

$\overrightarrow{AC} = 18\mathbf{i} - 3\mathbf{j}$ km **[1]**

b) $\overrightarrow{AC} = 18\mathbf{i} - 3\mathbf{j}$ km

So distance AC is $\sqrt{18^2 + (-3)^2}$ **[1]**

$= 3\sqrt{37}$ km **[1]**

Vectors 2

QUICK TEST (page 91)

1. a) $\begin{pmatrix} 0 \\ 7 \end{pmatrix}$ b) Magnitude $\frac{5\sqrt{5}}{2}$, on a bearing of 333.4°

c) $-4\mathbf{i} + 7\mathbf{j}$

2. a) $\mathbf{a} + \mathbf{b}$ b) $-2\mathbf{a}$ c) $-3\mathbf{a} + 2\mathbf{b}$

d) $3\mathbf{a} - \frac{3}{2}\mathbf{b}$ e) $k\left(3\mathbf{a} - \frac{3}{2}\mathbf{b}\right)$ f) $-2\mathbf{a} + 2\mathbf{b}$

3. a) $\overrightarrow{BA} = 15\mathbf{i} - 14\mathbf{j}$ b) $\overrightarrow{AB} = -15\mathbf{i} + 14\mathbf{j}$

PRACTICE QUESTIONS (page 91)

1. a) $\overrightarrow{AB} = \overrightarrow{AD} + \overrightarrow{DC} + \overrightarrow{CB}$

$\overrightarrow{CB} = -\frac{1}{3}\overrightarrow{AD}$ **[1]**

$\overrightarrow{AB} = \frac{2}{3}\mathbf{p} + \mathbf{q}$ **[1]**

b) Collinear so $\overrightarrow{DE} = k\,\overrightarrow{DC}$ **[1]**

$\overrightarrow{AE} = \overrightarrow{AD} + \overrightarrow{DE} = \mathbf{p} + k\mathbf{q}$ **[1]**

c)

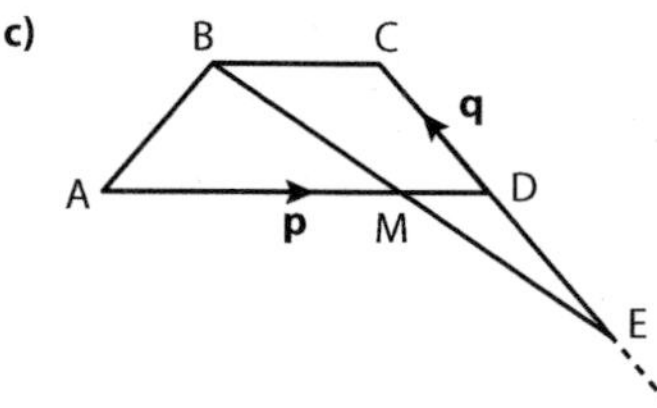

$\overrightarrow{BM} = \overrightarrow{BC} + \overrightarrow{CD} + \overrightarrow{DM} = \frac{1}{3}\mathbf{p} - \mathbf{q} - \frac{1}{7}\mathbf{p}$

$= \frac{4}{21}\mathbf{p} - \mathbf{q}$ **[1]**

Given that collinear with B and M

$\overrightarrow{BE} = k\left(\frac{4}{21}\mathbf{p} - \mathbf{q}\right)$ **[1]**

Alternate route $\overrightarrow{BE} = \overrightarrow{BC} + \overrightarrow{CE}$ **[1]**

$\overrightarrow{BE} = \frac{1}{3}\mathbf{p} - c\mathbf{q}$ **[1]**

Equate coefficients **p**

$\frac{4k}{21} = \frac{1}{3}$ **[1]**

$k = \frac{7}{4}$ **[1]**

Substitute k back into equation

$\overrightarrow{BE} = \frac{1}{3}\mathbf{p} - \frac{7}{4}\mathbf{q}$ **[1]**

2. a) $\overrightarrow{AB} = \overrightarrow{OB} - \overrightarrow{OA} = p\mathbf{i} - (q\mathbf{i} + 12\mathbf{j})$ **[1]**

$= (p - q)\mathbf{i} - 12\mathbf{j}$ km **[1]**

b) Use Pythagoras' Theorem for OA

$37^2 = q^2 + 12^2$ **[1]**

$q^2 = 1225 \rightarrow q = 35$ (ignore negative as p is positive) **[1]**

Use trigonometry with given bearing to find value of $(p - q)$

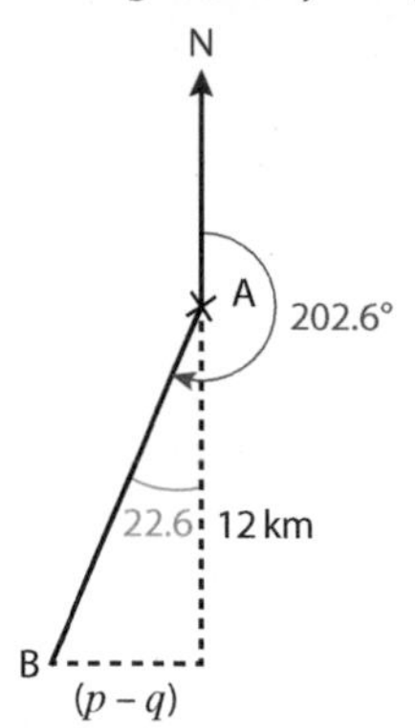

$\frac{p-q}{12} = \tan 22.6$ **[1]**

$p - 35 = 12 \tan 22.6 = 4.99511 \ldots$ **[1]**

$p = 30$, as it is an integer as specified by question, or $p - q = 5$

$AB = \sqrt{12^2 + 5^2} = \sqrt{169} = 13$ **[1]**

Day 6

Kinematics 2

QUICK TEST (page 95)

1. a) $v^2 = u^2 + 2as$ b) $v = u + at$
   c) $s = ut + \frac{1}{2}at^2$ d) $s = vt - \frac{1}{2}at^2$
   e) $s = \frac{1}{2}(u + v)t$
2. a) $-2\ ms^{-1}$ b) $1\ ms^{-2}$ c) 24 m
3. a) $v = 2t^3 - 4t$ b) $8\ ms^{-1}$
4. a) $s = \frac{1}{2}(u + v)t$

   $91 = \frac{1}{2}(26)t \rightarrow t = 7\ s$

   b) $v^2 = u^2 + 2as$

   $a = \frac{0 - 26^2}{2 \times 91} = -\frac{26}{7} = -3.71\ ms^{-2}$

   $\therefore$ deceleration of $3.71\ ms^{-2}$

PRACTICE QUESTIONS (page 95)

1. a) $v = \int a\ dt = \frac{3}{2}t^2 + 2t + C$ **[2 marks: lose 1 mark for each incorrect term]**

   b) $r = \int v\ dt = \frac{1}{2}t^3 + t^2 + Ct + k$ **[3 marks: lose 1 mark for each incorrect term]**

   c) $3.5 = \frac{1}{2} \times 1^3 + 1^2 + C + k$

   $C + k = 2$ **[1]**

   $13 = \frac{1}{2} \times 2^3 + 2^2 + C \times 2 + k$

   $2C + k = 5$ **[1]**

   $C = 3$ and $k = -1$ **[1]**

   $r = \frac{1}{2}t^3 + t^2 + 3t - 1 = \frac{1}{2} \times 3^3 + 3^2 + 3 \times 3 - 1$ **[1]**

   $= 30.5\ m$ **[1]**

2. a) Identifies the following values from the question

   $u = 6\ cms^{-1} = 0.06\ ms^{-1}$, $s = 50\ cm = 0.5\ m$, $a = -0.2\ cms^{-2}$
   $= -0.002\ ms^{-2}$ **[1]**

   Either uses $v = 0$ and shows that $s > 50\ cm$ or finds v and shows $v > 0$

   $v^2 = u^2 + 2as$

   $v^2 = 6^2 + 2 \times -0.2 \times 50 = 16$ **[1]**

   (If $v^2 > 0$ then the marble will reach the edge of the table)

   $v = 4\ cms^{-1}$ so the marble will reach the edge of the table (travelling at $0.04\ ms^{-1}$) **[1]**

   b) Vertically (downwards positive, measures in seconds and centimetres):

   $s = 0.75$, $u = 0$, $v =$ irrel., $a = 9.8$, $t = ?$

   $s = ut + \frac{1}{2}at^2$ **[1]**

   $0.75 = 0 + \frac{1}{2} \times 9.8 \times t^2$

   $t^2 = \frac{15}{98}$ **[1]**

   $t = \frac{\sqrt{30}}{14}$ (take positive, as negative not meaningful in this context) **[1]**

   Horizontally, velocity having left the table = $4\ cms^{-1}$ **[1]**

   Horizontal distance from edge of table to where marble hits the floor = vt

   $= 4 \times \frac{\sqrt{30}}{14}$ **[1]**

   $= 1.56\ cm$ (3 s.f.) **[1]**

   c) The marble has been modelled as a particle. This means it doesn't experience air resistance as it falls from the table. The acceleration due to gravity has been modelled as constant. Given the change in height of the marble is relatively small, this is a fair assumption to make. **[1 mark for any example of a modelling assumption with an explanation about it.]**

Forces

QUICK TEST (page 101)

1. a) 12 N to the right b) 3 N to the right, $\begin{pmatrix} 3 \\ 0 \end{pmatrix}$ c) 0

   d) $-6\mathbf{i} - 8\mathbf{j}$ (magnitude 10 N at 143.1° (1 d.p.) clockwise to the **i** direction)
2. a) $4\ ms^{-2}$ b) $75\ ms^{-2}$
   c) The body is in equilibrium. $0\ ms^{-2}$
   d) $-12\mathbf{i} - 16\mathbf{j}\ ms^{-2}$, $20\ ms^{-2}$ at 143.1° (1 d.p.) clockwise to the **i** direction
3.

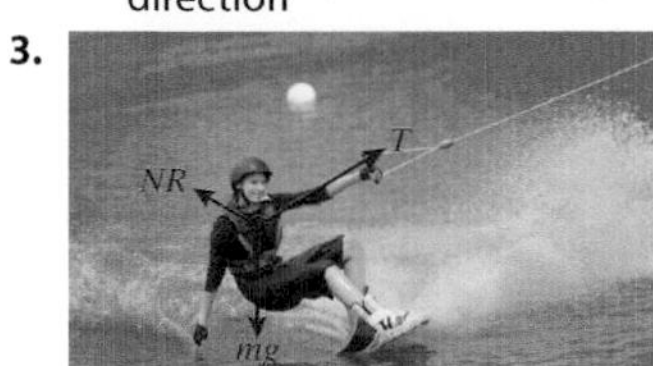

4. a)

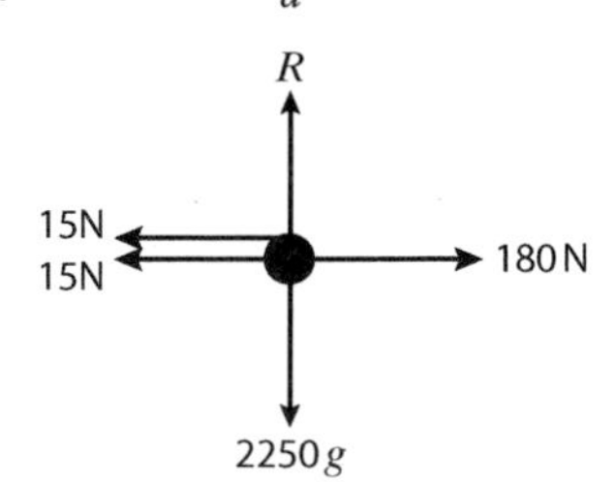

   b) $a = \frac{150}{2250} = \frac{1}{15} = 0.067\ ms^{-2}$ (3 d.p.)

   c)

   a

   $R = 750g$

   15 N ← → T

   $750g$

   a

   $R = 1500g$

   T ← 15 N → 180 N

   $1500g$

d) Considering the caravan and using $F = ma$:

Resultant force $= ma$

$T - 15 = 750 \times \frac{1}{15}$

$T = 50 + 15$

$T = 65$ N

PRACTICE QUESTIONS (page 101)

1. a) Total mass of car and trailer = 2022 kg

$F = ma$

Driving force (P) – combined resistance forces

$= 2022 \times 2 = 4044$ **[1]**

$P - 550 = 4044$ **[1]**

$P = 4594$ N **[1]**

b) For section AB:

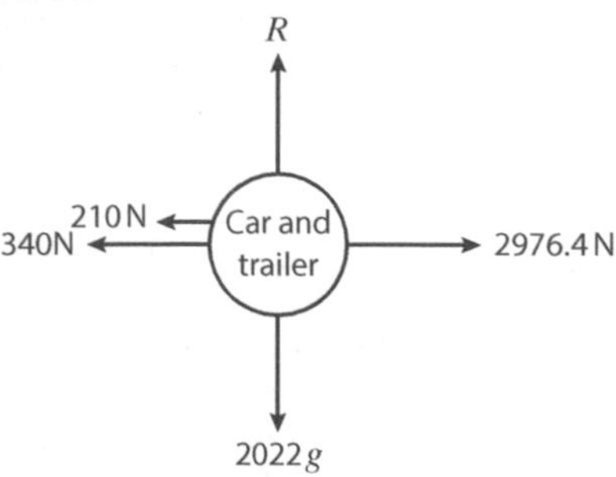

$F = ma$

$2976.4 - 550 = 2022a$ **[1]**

$a = \frac{6}{5} = 1.2$ ms^{-2} **[1]**

$s =$ irrel., $u = 12$, $v = ?$, $a = 1.2$, $t = 10$

$v = u + at$ **[1]**

$v = 12 + 1.2 \times 10 = 24$ ms^{-1} **[1]**

For trailer in section BC:

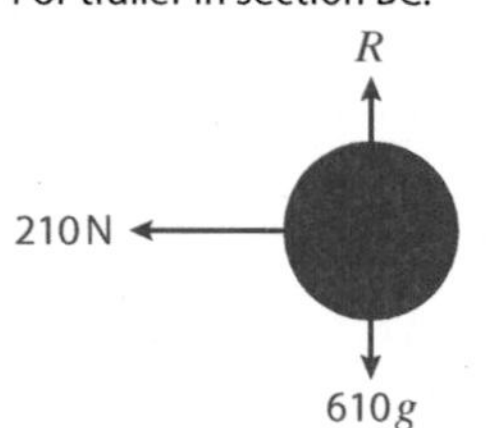

$F = ma$

$-210 = 610a$ **[1]**

$a = -\frac{21}{61}$ **[1]**

$s = ?$, $u = 24$, $v = 0$, $a = -\frac{21}{61}$, $t =$ irrel.

$v^2 = u^2 + 2as$

$0 = 24^2 + 2 \times \left(-\frac{21}{61}\right) \times s$ **[1]**

$s = \frac{24^2}{2 \times \left(\frac{21}{61}\right)} = \frac{5856}{7} = 836.6$ m (1 d.p.) **[1]**

c) For car in section BP:

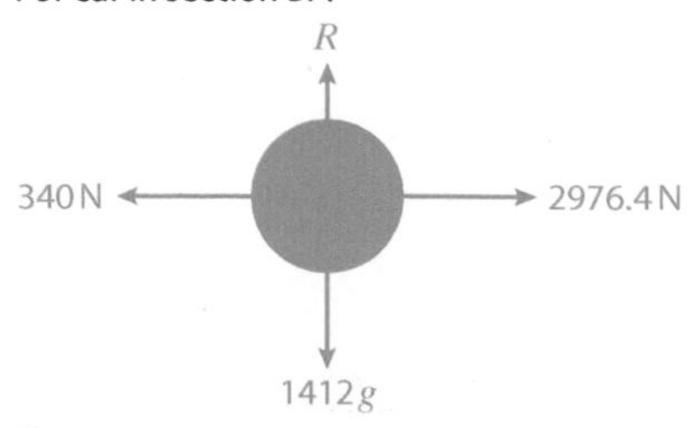

$F = ma$

$2976.4 - 340 = 1412 \times a$ **[1]**

$a = 1.86713 \ldots$ **[1]**

$s = \frac{836.6}{4} = 209.15$, $u = 24$, $v = ?$, $a = 1.86713\ldots$, $t =$ irrel.

$v^2 = u^2 + 2as$

$v^2 = 24^2 + 2 \times (1.86713 \ldots) \times 209.15 = 1357.0241 \ldots$ **[1]**

$v = 36.83 \ldots$ **[1]**

For car in section PC:

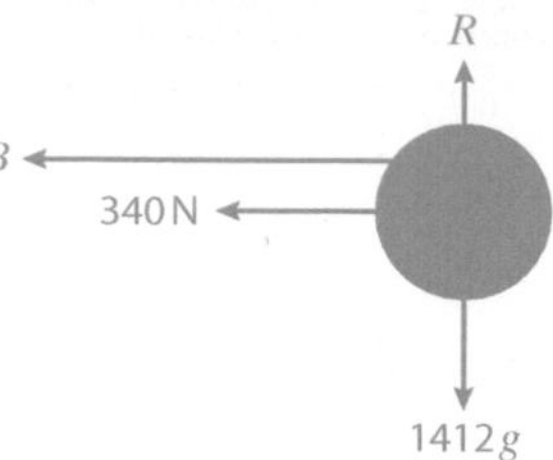

$s = 209.15 \times 3 = 627.45$, $u = 36.83 \ldots$, $v = 0$, $a = ?$, $t =$ irrel.

$a = \frac{v^2 - u^2}{2s}$

$a = \frac{-36.8378\ldots^2}{2 \times 627.45} = -1.08138032\ldots$ **[1]**

$F = ma$

$-B - 340 = 1412 \times -1.0813 \ldots$ **[1]**

$B = 1186.8 \ldots = 1190$ N (3 s. f.) **[1]**

2. a) For modelling situation as greatest imbalance of weights occur when the lift is either full or empty. Since $a = \frac{F}{m}$ the acceleration will be greatest for maximum resultant force and minimum mass. So consider the situation where the lift is empty. **[1]**

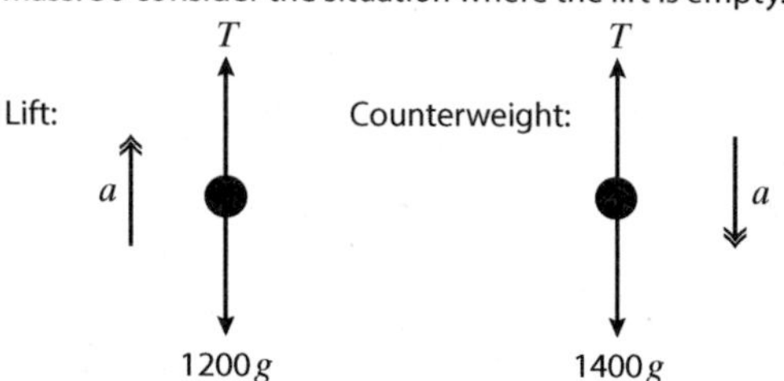

For force diagram **[1]**

Lift: resultant force $= T - 1200g$

For resultant force **[1]**

$1200a = T - 1200g$

Counterweight: $1400g - T = 1400a$

For use of $F = ma$ **[1]**

$T = 1400(g - a)$

$1200a = 1400(g - a) - 1200g$

$= 200g - 1400a$

For solving simultaneous equations **[2]**

$2600a = 200g$

$a = \frac{g}{13}$ ms^{-2} **[1]**

b) $T = 1400(g - a) = 1400\left(\frac{13g}{14}\right) = 1300g$ **[2]**

Proof

QUICK TEST (page 105)

1. a) Integers: 2, 3, 6, 7, 8, 9, 12, 16

b) Rational numbers: 2, 3, 6, 7, 8, $8.\dot{3}$, 9, 9.1, 12, $\frac{62}{5}$, 16

c) Square numbers: 9, 16

d) Prime numbers: 2, 3, 7

e) Irrational numbers: $\sqrt{2}$, π

2. 1×70, 2×35, 5×14, 7×10

3. Any three negative integers, e.g. –1, –2, –3

4. $n \in \mathbb{Z}$

$n, n + 1, n + 2$

5. Any example including at least one negative value, e.g. $4 + (-2) = 2$, as $2 < 4$ the statement is incorrect.

6. a) Anand is incorrect as 2 is a prime number, having factors only of 1 and itself. As 2 is an even number, it proves by counter example that not all primes are odd numbers.

b) All the primes between 10 and 50 are: 11, 13, 17, 19, 23, 29, 31, 37, 41, 43, 47.
For a number to be even it must end in a 0, 2, 4, 6 or 8. By exhaustion none of these numbers is prime so Anand is correct.
Alternatively, as no method is specified, you could use a deductive proof and say:
The only even prime is 2 as the definition of a prime number is that it has only one factor pair. $2 = 1 \times 2$
All other even numbers, n, have a factor pair of $2 \times m$, as well as $1 \times n$. So other primes must be odd.

7. Let $n \in Z^+$
A square number is n^2, the consecutive (next) square number is $(n+1)^2 = n^2 + 2n + 1$
$n^2 + n^2 + 2n + 1 = 2n^2 + 2n + 1$
$= 2(n^2 + n) + 1 = 2m + 1$ which is always odd since $2m$ is even.

PRACTICE QUESTIONS (page 105)

1. $(an+1)^2 = a^2n^2 + 2an + 1$ **[1]**
$(an-1)^2 = a^2n^2 - 2an + 1$ **[1]**
$(an+1)^2 - (an-1)^2 = (a^2n^2 + 2an + 1) - (a^2n^2 - 2an + 1)$
$= 4an$
$= 2 \times (2an)$ **[1]**
This is a multiple of 2. **[1]**

2. This question is asking for proof within a statistical context. As there are only six possible arrangements, it is possible to use proof by exhaustion to demonstrate all the possible combinations and draw the correct conclusion.

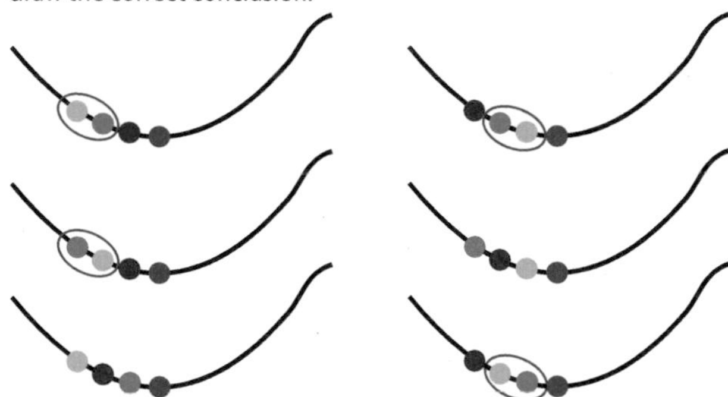

There are six possible arrangements of the four beads
(This must be drawn or written to demonstrate) **[1]**
There are four combinations where the blue and yellow are next to each other **[1]**

$\text{P(Blue next to yellow)} = \frac{\text{Number where YB next to each other}}{\text{Possible number of combinations}} = \frac{4}{6} = \frac{2}{3}$ **[1]**

3. a) One counter-example given to disprove statement, including clear demonstration that m in the given case is not prime.
$70 = 2 \times 35$, 35 isn't prime as $35 = 7 \times 5$. (To be prime there can only be a factor pair of $1 \times m$.) **[2]**
Conclusion: Billie is incorrect as $70 = 2 \times 5 \times 7$ **[1]**

b) Proof by exhaustion used **[1]**
Each step fully supported, no logic gaps **[1]**
Correct conclusion drawn **[1]**
$q \times 2$
$2 \times 2 = 4 = 2^2$ so 2 is half of a square number
$8 \times 2 = 16 = 4^2$ so 8 is half of a square number
$18 \times 2 = 36 = 6^2$ so 18 is half of a square number
$50 \times 2 = 100 = 10^2$ so 50 is half of a square number
$60.5 \times 2 = 121 = 11^2$ so 60.5 is half of a square number
$98 \times 2 = 196 = 14^2$ so 98 is half of a square number
Billie is correct as each number in the set is half of a square number.

Measures of Location and Spread

QUICK TEST (page 109)

1. $\bar{x} = 0.445$ (3 s.f.), $\sigma^2 = 0.0149$ (3 s.f.)
2. a) 35–39 mm **b)** $\bar{x} = 42.825$ mm
3. $4 - 2 = 2$
4. 80.1 (3 s.f.)
5. $13 - 11 = 2$
6. 29
7. Use median and IQR as females have an outlier
Female $Q_2 = 59$, IQR $= 63 - 52 = 11$
Male $Q_2 = 57$, IQR $= 62 - 56 = 6$
Females scored higher overall as median is higher, although males have a smaller IQR showing their results are more consistent.
8. $\bar{q} = 32, \sigma_q = 15$

PRACTICE QUESTIONS (page 109)

1. a) Input midpoints and frequencies into the statistical function on a calculator.
1997:
$\bar{x} = 179$ (3 s.f.) **[1]**
$\sigma = 37.6$ (3 s.f.) **[1]**
2017:
$\bar{x} = 201$ **[1]**
$\sigma = 35.5$ (3 s.f.) **[1]**

b) Songs are on average longer in 2017 than in 1997 as the mean is higher **[1]**. 2017 also has a smaller standard deviation, meaning songs are of a more consistent length **[1]**.

2. a) $\bar{l} = \frac{\sum l}{n} = \frac{1608}{60} = 26.8$ **[1]**
$\sigma = \sqrt{\frac{\sum l^2}{n} - \left(\frac{\sum l}{n}\right)^2} = \sqrt{\frac{53129}{60} - \left(\frac{1608}{60}\right)^2}$ **[1]**
$= 12.9$ (3 s.f.) **[1]**

b) It will go down **[1]** as she produced above the average amount of milk **[1]**.

c) Subtract 32 from $\sum l$, and 32^2 from $\sum l^2$
New $\sum l = 1576$ **[1]**
New $\sum l^2 = 52\,105$ **[1]**

d) $\bar{y} = \frac{1080}{30} = 36$ **[1]**
$\bar{l} = \frac{\bar{y}}{4} - 3 = 6$ **[1]**
$\sigma_y = \sqrt{\frac{41\,024}{30} - \left(\frac{1080}{30}\right)^2} = 8.453795\ldots$ **[2]**
$\sigma_l = \frac{\sigma_y}{4} = 2.11$ (3 s.f.) **[1]**
The mean and standard deviation of y are calculated in the usual way and then uncoded by changing the subject of the coding formula, so $\frac{\bar{y}}{4} - 3 = \bar{l}$.

Day 7

Probability

QUICK TEST (page 112)

1. $P(A \cap B) = 0.06$
2. $P(D) = 0.3$
3. a) 0.6 **b)** 0.5 **c)** 0.2 **d)** 0.9
4. $\frac{7}{15}$
5.

Spinner 2 \ Spinner 1 (+)	1	3	5	7
2	3	5	7	9
4	5	7	9	11
6	7	9	11	13
8	9	11	13	15

$P(\leqslant 7) = \frac{6}{16} = \frac{3}{8}$

6. $\frac{18}{135} = \frac{2}{15}$, the dice looks biased towards a 3, therefore not fair.

7.

M B
6 14 3
7

PRACTICE QUESTIONS (page 113)

1. a) Factory F = broken

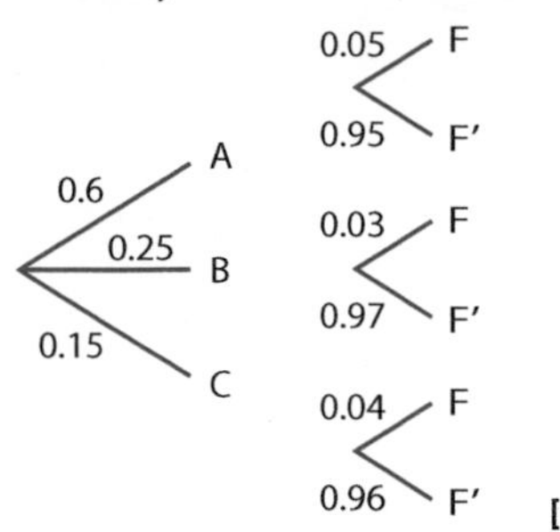

[3]

b) i) 0.25×0.03 [1]
$= 0.0075$ [1]

ii) $(0.6 \times 0.95) + (0.25 \times 0.97) + (0.15 \times 0.96)$ [1]
$= 0.9565$ [1]

2. a) P and C or R and C [1]

b) $\frac{5}{70} = \frac{1}{14}$ (1)

c) $P(P \cap R) = 0.0714$ (4 d.p.) from Venn [1]
$P(P \cap R) = P(P) \times P(R)$ using multiplication rule
$= 0.1122$ (4 d.p.) [1]
P and R are not independent as the multiplication rule does not give the intersection [1]
Note: Use of the notation is not required but is accepted.

3. a) $\frac{65 + 50 + 72}{285}$ [1]
$= \frac{187}{285}$ [1]

b) $2\left(\frac{115}{285} \times \frac{72}{284}\right)$ [1]
$= \frac{276}{1349}$ [1] $= 0.205$ (3 s.f.) [1]

Statistical Sampling

QUICK TEST (page 115)

1. A census uses all of the population, whereas a sample is part of it.
2. Opportunity/convenience sampling. Will be biased; students going into the library are more likely to read than other students.
3. Number all employees from 1 to 360. $360 \div 30 = 12$. Randomly choose a starting point between 1 and 12 using a random number generator then choose every 12th employee.
4. Quota sampling; quick and easy but not representative of population.
5. Boys: $\frac{30}{260} \times 160 = 18.46$; Girls: $\frac{30}{260} \times 100 = 11.54$
Use a random sampling method to choose 18 boys and 12 girls.

Representing Data

QUICK TEST (page 119)

1. $\text{IQR} = Q_3 - Q_1 = 23 - 18 = 5$
$1.5 \times \text{IQR} = 7.5$
$Q_1 - 1.5 \times \text{IQR} = 10.5$, therefore no low outliers
$Q_3 + 1.5 \times \text{IQR} = 30.5$, therefore 38 is an outlier
2. a) The runners' best time decreases by 0.85 seconds every year.
b) Not suitable. Starting time outside the range of original data therefore extrapolation and unreliable. Also study was on adults; model may be different for children.

3. Positive correlation – the older the car the more miles it has done.
4. Median = 54 mins
IQR = 68 – 38 = 30

PRACTICE QUESTIONS (page 119)

1. a) No. 50% scored above 65% in the test as 65 is the median. [1]
b) An outlier [1]
c) 40 [1]
Since 40 is the lower quartile and 75% of the data lies above the lower quartile [1]
d) Girls performed better on average as their median (65) is higher than the boys' median (50) [1]
Boys' results were more consistent as their IQR (25) is smaller than the girls' (30) [1]
2. a) 700 small squares in bars between 65 and 80 represent 35 sunflowers
$\frac{35}{700} = 0.05$ sunflowers per square [1]
Between 40 and 60 there are 600 small squares [1]
$0.05 \times 600 = 30$ [1]
b) 90 altogether

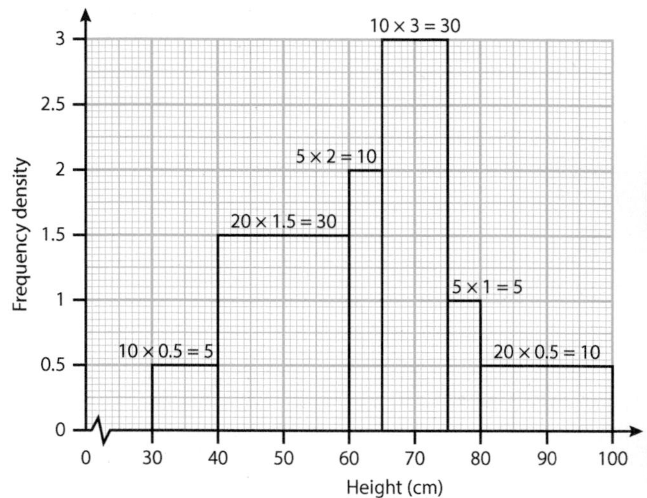

[1 mark for calculations; 1 mark for answer]

Statistical Distributions

QUICK TEST (page 123)

1. a) No, distance is continuous
b) Yes, number of heads is discrete
2. $c = \frac{1}{6}$
3. $k = \frac{5}{12}$ $P(R > 6) = \frac{2}{3}$
4. 0.0881
5. a) $P(X \leqslant 7) = 0.9302$
b) $P(X < 4) = P(X \leqslant 3) = 0.3217$
c) $P(X > 10) = 1 - P(X \leqslant 10) = 0.0029$
d) $P(X \geqslant 8) = 1 - P(X \leqslant 7) = 0.0698$
6. $y = 3$
7. $P(3 \leqslant M < 8) = P(M \leqslant 7) - P(M \leqslant 2) = 0.1531$
8. $\frac{9}{20}$
9. $k = \frac{1}{94}$
10. $X \sim B(20, 0.5)$
$P(X \geqslant 8) = 1 - P(X \leqslant 7) = 0.8684$

PRACTICE QUESTIONS (page 123)

1. a) First write the distribution, $X \sim B(50, 0.7)$
   Identify the probability to be found (use calculator function as $p > 0.5$)
   $P(X > 36) = 1 - P(X \leqslant 36)$ **[1]**
   $= 0.3279$ **[1]**
   b) No **[1]**, a different town may not have the same success **[1]**.
2. a) Substitute y-values into each part to give $\frac{1}{4}+\frac{2}{4}+3k+4k=1$ **[1]**
   $k = \frac{1}{28}$ **[1]**
   b)

y	1	2	3	4
$P(Y = y)$	$\frac{1}{4}$	$\frac{2}{4}$	$\frac{3}{28}$	$\frac{4}{28}$

   Table set up correctly **[1]**
   Probabilities in correct places (no k) **[1]**
   c) $P(2 \leqslant Y < 4) = P(Y = 2) + P(Y = 3)$
   $= \frac{17}{28}$ **[1]**
3. a) List all of the people who took a test in the last year from that driving school and number them. Take a simple random sample of 30 using a random number generator (or names in a hat). Systematic sampling also a valid answer. **[1]**
   b) Looking at fails so $F \sim B(30, 0.05)$ **[1]**
   Find $P(F < 3) = P(F \leqslant 2) = 0.8122$ **[1]**
   c) $P(F > f) = 0.0156$
   $P(F > 4) = 1 - P(F \leqslant 4) = 0.0156$ **[1]**
   Therefore more than four people must fail from that driving school for it to be investigated. **[1]**

Hypothesis Testing

QUICK TEST (page 127)

1. $H_0 : p = \frac{1}{2}$, $H_1 : p \neq \frac{1}{2}$
2. $X \leqslant 3$
3. $P(X \leqslant 1) = 0.085$, accept H_0 as $0.085 > 0.05$
4. $P(X \leqslant 4) = 0.0023$
   $P(X \leqslant 5) = 0.0086$
   $P(X \geqslant 18) = 1 - P(X \leqslant 17) = 0.0058$
   $P(X \geqslant 19) = 1 - P(X \leqslant 18) = 0.0016$
   The critical region is $X \leqslant 4$ and $X \geqslant 19$
5. $P(X \geqslant 8) = 1 - P(X \leqslant 7) = 0.0048$, reject H_0 as $0.0048 < 0.02$, and conclude it is likely that $p > 0.35$
6. A hypothesis test is where an assumption regarding a population parameter is tested against an alternative.
7. $P(Y \leqslant 3) = 0.0057$
   $P(Y \leqslant 4) = 0.0185$
   Answer 0.0057
8. $P(X \leqslant 8) = 0.0303$
   $P(X \leqslant 9) = 0.0644$
   $P(X \geqslant 19) = 1 - P(X \leqslant 18) = 0.0699$
   $P(X \geqslant 20) = 1 - P(X \leqslant 19) = 0.0363$
   The critical region is $X \leqslant 9$ and $X \geqslant 20$, actual significance level of test: $0.0644 + 0.0363 = 0.1007$

PRACTICE QUESTIONS (page 127)

1. $X \sim B(40, 0.3)$ **[1]**
   $H_0: p = 0.3$, $H_1: p < 0.3$ **[1]**
   $P(X \leqslant 6) = 0.0238$ **[1]**
   Reject H_0 as $0.0238 < 0.05$ **[1]**
   There is evidence at the 5% level to suggest it is likely that Hastozena's new drug works for less than 30% of the population. **[1]**
2. a) $X \sim B(50, 0.8)$ **[1]**
   $H_0: p = 0.8$, $H_1: p \neq 0.8$ **[1]**
   $np = 50 \times 0.8 = 40$, hence calculate $P(X \leqslant 35)$
   $P(X \leqslant 35) = 0.0607$ **[1]**
   Accept H_0 as $0.0607 > 0.05$ **[1]**
   There is no evidence at the 10% level to suggest the number of first goals scored in the second half of each hockey game has changed. **[1]**
   b) The goals are scored independently/the probability of scoring a goal remains constant. **[1]**
3. $X \sim B\left(20, \frac{1}{4}\right)$ **[1]**
   $H_0: p = \frac{1}{4}$, $H_1: p > \frac{1}{4}$ **[1]**
   $P(X \geqslant 10) = 1 - P(X \leqslant 9) = 0.0139$
   $P(X \geqslant 11) = 1 - P(X \leqslant 10) = 0.0039$
   Critical region is $X \geqslant 11$ **[1]**
   Accept H_0 as 9 is not in the critical region **[1]**
   There is no evidence at the 1% level to suggest the number of rose bushes with red flowers has increased. **[1]**
4. a) Two from: fixed number of trials; two outcomes – success or fail; constant probability; independent probabilities. **[2]**
   b) $P(X \leqslant 11) = 0.0125$
   $P(X \leqslant 12) = 0.0253$ **[1]**
   $P(X \geqslant 28) = 1 - P(X \leqslant 27) = 0.0342$
   $P(X \geqslant 29) = 1 - P(X \leqslant 28) = 0.02002$ **[1]**
   The critical region is $X \leqslant 12$ and $X \geqslant 29$ **[1]**
   c) Actual significance level of test is $0.0253 + 0.02002 = 0.04532$ **[2]**

Index

acceleration 75, 78–83, 92–95
adding vectors 88–90
air resistance 94
asymptotes 20
averages 106–109

binomial distribution 121–122
binomial expansion 28–31
brackets 8–9, 28–30

calculus 93–95
circles 26
class boundaries 106
coefficients 9, 28–31
collinearity 89–90
column notation 85
completing the square 10–11
components 84–87
composite areas 70–72
coordinates 24–27, 85
cosine rule 33

data representation 116–119
differentiation 56–63
dimensional analysis 76
discrete distributions 120–121
discriminant 11, 21
displacement–time graphs 78–83
distributions 120–123

elimination 15
equilibrium 97–98
expanding brackets 8–9, 28–30
exponentials 21, 40–43, 54

factorising 9, 10
forces 96–101
frequency polygons 116
friction 96

geometrical problem solving 89
gradients 56–58
graphs 20–23, 34, 40–47, 78–83, 104
gravity 93–94

histograms 117–118
hypothesis testing 124–127

indices 4–7
inequalities 16–19
integration 64–73

kinematics 78–83, 92–95

linear functions 20
logarithms 44–55

magnitudes 85
mass 75, 93–94
mean, median and mode 106–108
measures of location 106–108
mechanics 74–101
modelling 74–77, 92–95
motion 75, 78–83, 92–95, 98–99

normal 61–62
normal reaction 96

outliers 117

particles 94, 99–100
polynomials 8–13, 20–23, 52–54, 57–58
position vectors 90
prime numbers 102–103
probability 110–113
projectiles 94
proof 102–105
proportionality 21

quadratics 9–13, 17, 20
quantities 74–75
quartiles 107

range 107–108
reciprocal curves 21
roots 20

scalars 84, 88
scatter diagrams 118
set notation 102
simple equations 14
simultaneous equations 14–19
standard deviation 107–108
stationary points 60–61
statistical sampling 114–115
substitution 16
surds 5–7

tangents 56–58, 61–62
tension 96
thrust 96
transformations 22–23
tree diagrams 110
trigonometry 32–38
turning points 20
types of number 102–103

units 74–75

vectors 84–91
velocity–time graphs 79–83
Venn diagrams 111

weight 75, 93–94

Acknowledgements

The authors and publisher are grateful to the copyright holders for permission to use quoted materials and images. Every effort has been made to trace copyright holders and obtain their permission for the use of copyright material. The authors and publisher will gladly receive information enabling them to rectify any error or omission in subsequent editions. All facts are correct at time of going to press.

Published by Letts Educational
An imprint of HarperCollins*Publishers*
1 London Bridge Street
London SE1 9GF

ISBN: 9780008276034

First published 2019

10 9 8 7 6 5 4 3 2 1

British Library Cataloguing in Publication Data.

A CIP record of this book is available from the British Library.

Authors: Rosie Benton and Sharon Faulkner
Commissioning Editor: Gillian Bowman
Project Manager: Richard Toms
Editorial: Jess White and Richard Toms
Indexer: Simon Yapp

Inside Concept Design: Ian Wrigley
Cover Design: Sarah Duxbury
Text Layout: QBS Learning
Production: Karen Nulty
Printed and bound in Italy by Grafica Veneta S.p.a